Heat Exchangers

Performance Prediction & Evaluation

by D. James Benton

Copyright © 2017 by D. James Benton, all rights reserved.

Foreword

Heat is that transient form of energy that passes through a system boundary by virtue of a temperature difference. This temperature difference distinguishes heat from *work*, which passes through a system boundary by virtue of a force. Systems may contain energy, but do *not* contain heat or work.

We are most often concerned with heat transfer within an enclosed device or heat exchanger. There are several ways to analyze the heat transfer within a heat exchanger, all of which involve assumptions. Often these assumptions are unstated, yet they impact the outcome of the analysis. These methods and assumptions will be presented along with examples of when the assumptions work and fail.

The two areas of focus in heat exchanger analysis are performance prediction and evaluation, that is: design and testing. Thermal design will be presented here, but mechanical design will not. Actual test data will be presented along with uncertainty analysis, as this is critical to a comprehensive understanding of heat exchanger performance.

This book is divided into two sections: Chapters 1 through 10 cover classical (i.e., analytical) methods and Chapters 11 and following cover numerical methods. Prediction and evaluation are included in both sections. Some sections of code are included in the Appendices. The complete codes and precompiled executables are included in the on-line archive.

All of the examples contained in this book,
(as well as a lot of free programs) are available at...
https://www.dudleybenton.altervista.org/software/index.html
Example spreadsheets are provided
in both SI and English units.

Table of Contents

	page
Foreword	i
Chapter 1. Log-Mean Temperature Difference	1
Chapter 2. The NTU-Effectiveness Method	3
Chapter 3. The F-LMTD Method	5
Chapter 4. Water-Cooled Steam Surface Condenser	11
Chapter 5. Air-Cooled Steam Surface Condenser	19
Chapter 6. Simple Shell-and-Tube Heat Exchanger	27
Chapter 7. Feedwater Heater	33
Chapter 8. Heat Recovery Steam Generator	37
Chapter 9. Moisture Separator/Reheater	45
Chapter 10. The TEMA Designs	49
Chapter 11. Simple Numerical Methods	53
Chapter 12: Variable Properties	59
Chapter 13: Variable Conductance	63
Chapter 14. Two-Phase Flow Inside Tubes	67
Chapter 15. Condensation in Crossflow	75
Chapter 16. Operational Data	81
Chapter 17. Monte Carlo Methods	87
Appendix A. Crossflow Program	93
Appendix B. Moisture Separator/Reheater Program	97
Appendix C: Monte Carlo Codes	107

Chapter 1. Log-Mean Temperature Difference

Unlike companion fields, such as thermodynamics and fluid flow, there are relatively few governing equations in heat transfer and even fewer in heat exchanger analysis. The fundamental one being:

$$dQ = U\Delta T dA \qquad (1.1)$$

The differential heat transfer rate (per unit area), dQ, is equal to the overall conductance (i.e., series combination of all heat transfer coefficients), U, times the local temperature difference, ΔT, times the differential area, dA. All of these variables may vary over the surface of the heat exchanger, thus integrating this differential equation requires different approaches. We will begin with the most simple.

The first consideration is geometry. The simplest geometry would be a double pipe, where one fluid flows through the inner pipe and a second fluid flows through the annulus formed between the two pipes. There are two flow configurations: co-current and counter-current. A second consideration is thermal properties, primarily specific heat. At this point, we will consider both fluids to have constant specific heats and use the notation, C_P.

The mass flow rates will be given the symbol, m, and the subscripts H and C will be used to designate the hot and cold streams, respectively. The conservation of energy leads to:

$$dQ = \dot{m}_C C_{PC} \, dT_C = -\dot{m}_H C_{PH} \, dT_H \qquad (1.2)$$

The negative sign before the third term in Equation 1.2 arises from the fact that as the temperature of the cold stream increases the temperature of the hot stream decreases. The temperature difference is given by:

$$\Delta T = T_H - T_C \qquad (1.3)$$

The differential of the temperature difference is given by:

$$d\Delta T = dT_H - dT_C \qquad (1.4)$$

Equation 1.2 can be rearranged to form:

$$dT_H - dT_C = -\frac{dQ}{\dot{m}_H C_{PH}} - \frac{dQ}{\dot{m}_C C_{PC}} \qquad (1.5)$$

Equations 1.4 and 1.5 can be substituted into Equation 1.1 and both sides divided by ΔT to yield:

$$\left(\frac{\dot{m}_H C_{PH} \dot{m}_C C_{PC}}{\dot{m}_H C_{PH} + \dot{m}_C C_{PC}} \right) \frac{d\Delta T}{\Delta T} = U dA \qquad (1.6)$$

Both sides of Equation 1.6 are easily integrated to arrive at:

$$\left(\frac{\dot{m}_H C_{PH} \dot{m}_C C_{PC}}{\dot{m}_H C_{PH} + \dot{m}_C C_{PC}}\right) \ln\left(\frac{\Delta T_{out}}{\Delta T_{in}}\right) = UA \quad (1.7)$$

The heat transfer is also equal to:

$$Q = \dot{m}_H C_{PH}\left(T_{H,in} - T_{H,out}\right) = \dot{m}_C C_{PC}\left(T_{C,out} - T_{C,in}\right) \quad (1.8)$$

These last two equations can be combined (substitute $mC_P=Q/\Delta T$ hot and cold from Eqn. 1.8 into Eqn. 1.7) to form:

$$Q = UA\frac{\Delta T_{out} - \Delta T_{in}}{\ln\left(\dfrac{\Delta T_{out}}{\Delta T_{in}}\right)} \quad (1.9)$$

The last group in Equation 1.9 is called the log-mean temperature difference or LMTD. Several cardinal rules of calculus were broken in the process of arriving at this result, including:

- U may depend on temperature, in which case, it should have been on the left side of Equation 1.6.
- The specific heats may not have been constant, in which case, they shouldn't have been taken outside the implied integral.
- ΔT might have been zero, in which case we shouldn't have divided by it when taking it to the left side of the differential equation.
- ΔT might not be zero, but dΔT might be. If $m_H C_{PH}=m_C C_{PC}$ and the configuration is counter-current, the difference in temperatures of the two streams is constant. Although dQ is not zero, by using Equation 1.5 we inadvertently multiplied both sides of Equation 1.1 by zero. The LMTD becomes 0/ln(1), which is undefined, but $\Delta T_{in}=\Delta T_{out}$.

The LMTD approach can fail in more complex arrangements with variable specific heats. Several examples will be given after other methods are introduced so that these may also be evaluated side-by-side. It is worth noting at this point that no method is adequate for every possible application and circumstance.

Know which approach is best
for your particular application
and what its limitations are.

Chapter 2. The NTU-Effectiveness Method

This method is similar to the LMTD method, but takes a slightly different approach to solving the same differential equation (viz., 1.1). The effectiveness, ε, is defined as the ratio of the actual to the maximum heat transfer:

$$\varepsilon = \frac{Q}{Q_{MAX}} \quad (2.1)$$

The heat transfer, Q, for both the hot and cold streams is given by Equation 1.8. The maximum possible heat transfer would be equal to the difference in the two inlet temperatures times the minimum product of the mass flow and specific heat or:

$$Q_{MAX} = (\dot{m}C_P)_{MIN}(T_{H,in} - T_{C,in}) \quad (2.2)$$

where the minimum could be the hot or cold side. The ratio of the minimum to maximum product of mass and specific heat is given the symbol, R:

$$R = \frac{(\dot{m}C_P)_{MIN}}{(\dot{m}C_P)_{MAX}} \quad (2.3)$$

In the special case of phase change (most often this occurs in a condenser) CP is infinite, making R=0. The number of transfer units, NTU, is defined as:

$$NTU = \frac{UA}{(\dot{m}C_P)_{MIN}} \quad (2.4)$$

Through the same steps used to arrive at Equations 1.5, 1.6, and 1.7, we can arrive at the following effectiveness for co-current flow in a double pipe:

$$\varepsilon = \frac{1 - e^{[-NTU(1+R)]}}{1 + R} \quad (2.5)$$

and for counter-current:

$$\varepsilon = \frac{1 - e^{[-NTU(1-R)]}}{1 - R e^{[-NTU(1-R)]}} \quad (2.6)$$

which reduces to the following when $R=1$:

$$\varepsilon = \frac{NTU}{1 + NTU} \quad (2.7)$$

Of course, the same cardinal rules of calculus were violated in arriving at these equations. Analytical solutions for many other heat exchanger geometries have been worked out and can readily be found on-line. These will only be covered here as they compare to more general numerical solutions. Computer solutions are so convenient and flexible that the analytical solutions have lost some of the utility they once had when computers were not as readily available.

The NTU-ε method leads to an even more useful approach by considering a more general definition of effectiveness:

$$P_I = \frac{q}{C_I(T_{hot,in} - T_{cold,in})} \quad (2.8)$$

where the subscript, I, can be either the hot or cold stream. The capacity of the stream is given by:

$$C_I = (\dot{m}C_P)_I \quad (2.9)$$

and the number of transfer units associated with stream, I, is given by:

$$NTU_I = \frac{UA}{(\dot{m}C_P)_I} \quad (2.10)$$

The capacitance ratio for the stream is given by:

$$R_I = \frac{C_I}{C_{II}} \quad (2.11)$$

where R_{II} is the non-reference (i.e., the other) stream. It also follows that:

$$P_I = \left(\frac{C_h}{C_c}\right)\left(\frac{T_{h,i} - T_{h,o}}{T_{h,i} - T_{c,i}}\right) = \left(\frac{C_c}{C_h}\right)\left(\frac{T_{c,o} - T_{c,i}}{T_{h,i} - T_{c,i}}\right) \quad (2.12)$$

Once cast in this form, the performance of many heat exchangers performance may be expressed analytically by P as a function of NTU_I, R, and the arrangement. Alternately, NTU_I may be expressed as a function of P, R, and the arrangement. The premier reference text on this subject is *Heat Transfer* by L. C. Thomas. There are several editions, but the Professional Edition is by far the most useful. One source to find this would be the following link:

https://www.amazon.com/Lindon-C.-Thomas/e/B001HOVPI4

This present text is merely a source of examples and illustrations. The reader is directed to Lindon's text for comprehensive details on the P-NTU$_I$ method and many other things, including: convection, radiation, boiling, condensation, and heat transfer coefficients.

Chapter 3. The F-LMTD Method

It should be evident that a double pipe cannot adequately approximate any but the most simplistic heat exchanger geometries. One approach has been to multiply by a fudge factor, F, to account for the geometry and still use the log-mean temperature difference. The F-LMTD method is somewhat useful, at least for simple geometries and constant properties plus it has some historical interest. Only the crossflow geometry will be considered here.

$$UA = \frac{Q}{F \times LMTD} \quad (3.1)$$

Performance Prediction

The method is most often presented in the form of parametric curves based on analytical solutions to the steady-state conduction equation. Separate analytical solutions must be obtained if either of the sides (i.e., hot/cold fluids) is mixed across the direction of flow. Before the advent of the microcomputer, these solutions were evaluated on mainframes and drawn on pen plotters—that held an actual pen and moved it or the paper around! The solution for both sides unmixed is illustrated in the following figure for values of R=0.2, 0.4, ... 4.0:

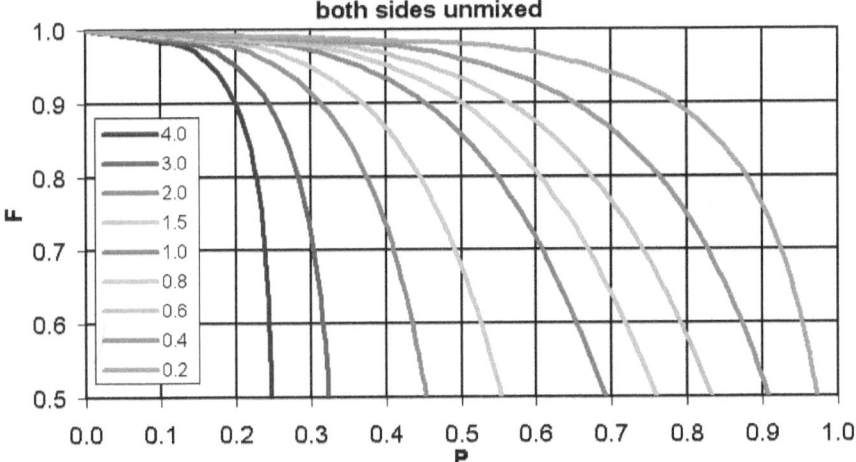

These crossflow plus similar analytical solutions may be cast in the form of $P_I(NTU_I, R_I,$ arrangement) and are strictly applicable in the case of constant properties. A variety of such solutions are provided in Lindon Thomas' book previously mentioned. Only one will be included here–that corresponding to the preceding figure (unmixed):

$$F = \frac{1}{NTU_I(1-R)} \ln\left[\frac{R\exp(-\Gamma R)}{R-1+\exp(-\Gamma R)}\right] \quad R < 1 \quad (3.2)$$

$$F = \frac{1}{NTU_I} \left[\frac{1-\exp(-\Gamma)}{\exp(-\Gamma)} \right] \quad R=1 \quad (3.3)$$

$$P = \sum_{m=0}^{\infty} \left\{ \frac{\Phi_m \Psi_m}{R\, NTU_I} \right\} \quad (3.4)$$

$$\Phi_m = 1 - \exp(-NTU_I) \sum_{j=0}^{m} \frac{NTU_I^{\,j}}{j!} \quad (3.5)$$

$$\Psi_m = 1 - R\exp(-R\, NTU_I) \sum_{j=0}^{m} \frac{(R\, NTU_I)^j}{j!} \quad (3.3)$$

$$P = \frac{1-\exp(-\Gamma R)}{R} \quad (3.7)$$

$$\Gamma = 1 - \exp(-NTU_I) \quad (3.8)$$

The crossflow solution for one side mixed is shown in the following figure:

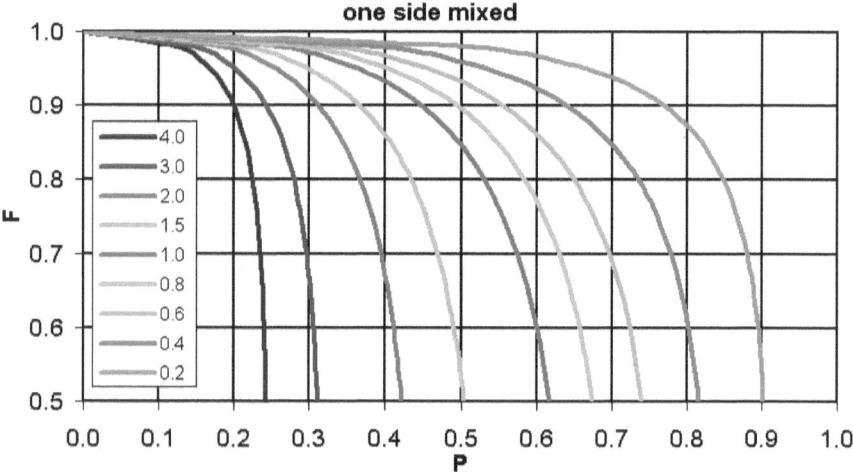

Notice that all of the curves in the second figure fall below and to the left of the corresponding curve in the preceding figure. For example, in the first figure (unmixed) the magenta $R=0.2$ curve intersects the horizontal axis ($F=0.5$) at $P=0.97$, while this same curve in the second figure (one side mixed) intersects at $P=0.90$. The both sides mixed solution yields even lower values of F, as shown in this next figure, where this same curve intersects at $P=0.89$. An examination of Equation 3.1 reveals that a smaller value of F will require a larger UA and more heat exchange surface, most likely costing more.

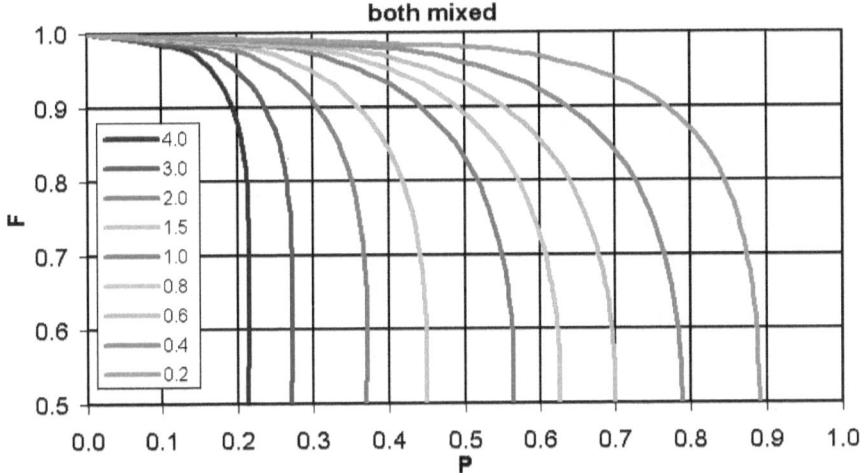

The F-LMTD method for all three crossflow configurations is illustrated in the Excel® spreadsheet named crossflow_analytical.xls, which is contained in the on-line archive. Macros (i.e., VBA® functions) are provided for all the preceding curves plus several more to facilitate their use in problem solving.

	A	B	C	D	E
1	crossflow example				
2	symbol	units	hot	cold	
3	U	W/m²/°C	150		
4	Cp	kJ/kg/°C	4.0	1.0	
5	m	kg/s	100	500	
6	Tin	°C	50	15	
7	Tout	°C	20	39.0	
8	q	kW	12,000		
9	LMTD	°C	7.61		
10	P	-	0.686		
11	R	-	0.800		
12	symbol	units	unmix	1mix	mixed
13	F	-	0.667	0.352	0.183
14	UA	kW/°C	2362	4481	8622
15	A	m²	15.7	29.9	57.5
16	user inputs in blue				
17	calculations in orange				

If any of the blue values are changed, the orange ones will automatically update, yielding the *F* factor and required surface area. In the case shown here, mixing either of the sides increases the required surface area from 15.7 to 29.9 m². Mixing both sides increases the required area to 57.5 m²! This example illustrates why mixing either of the two streams is not recommended from a thermal perspective, although from a maintenance or mechanical perspective it might be.

In general, it's not a good idea to mix either of the streams—think: Second Law of Thermodynamics. Mixing always results in the production of entropy, which is an irreversible process. Only mix the hot and/or cold sides in such a heat exchanger if there's some other reason, such as: to avoid fouling or plugging of small channels, to simplify fabrication, or to provide greater structural strength and eliminate the need for additional material.

Performance Evaluation

This unmixed design is presented as a performance test in the same spreadsheet. Test measurements include: the hot and cold inlet and outlet temperatures along with the hot and cold side flows. The uncertainty for each instrument is listed along with the average value over the test period. The instrument uncertainty is divided into two components: temporal and persistent.

Crossflow Heat Exchanger Test with Uncertainty

measurement	units	average	points	std.dev.	bias	total unc.	sensitivity	units	contrib
Thot,in	°C	51.25	30	0.82	0.14	0.34	-148.7	kW/°C²	-50
Thot,out	°C	20.85	30	0.79	0.14	0.33	-330.6	kW/°C²	-108
Tcold,in	°C	15.95	30	0.65	0.14	0.28	197.8	kW/°C²	56
Tcold,out	°C	40.35	30	0.63	0.14	0.28	281.6	kW/°C²	78
m,hot	kg/s	98.84	30	1.93%	2.0%	2.1%	55.41	kWs/kg/°C	116
m,cold	kg/s	497.55	30	1.62%	2.0%	2.1%	-6.156	kWs/kg/°C	-64
calculation	units	value							203
LMTD	°C	7.50							8.4%
q,hot	kW	12,019							
q,cold	kW	12,140							
q,avg	kW	12,080							
mCp,hot	kW/°C	395.358							
mCp,cold	kW/°C	497.549							
R	-	0.795							
P,hot	-	0.684							
P,cold	-	0.691							
F,hot	-	0.674							
F,cold	-	0.660							
UA,hot	kW/°C	2377							
UA,cold	kW/°C	2451							
UA,avg	kW/°C	2414							
UA,design	kW/°C	2362							
ΔUA	-	2.2%							

calculate sensitivities

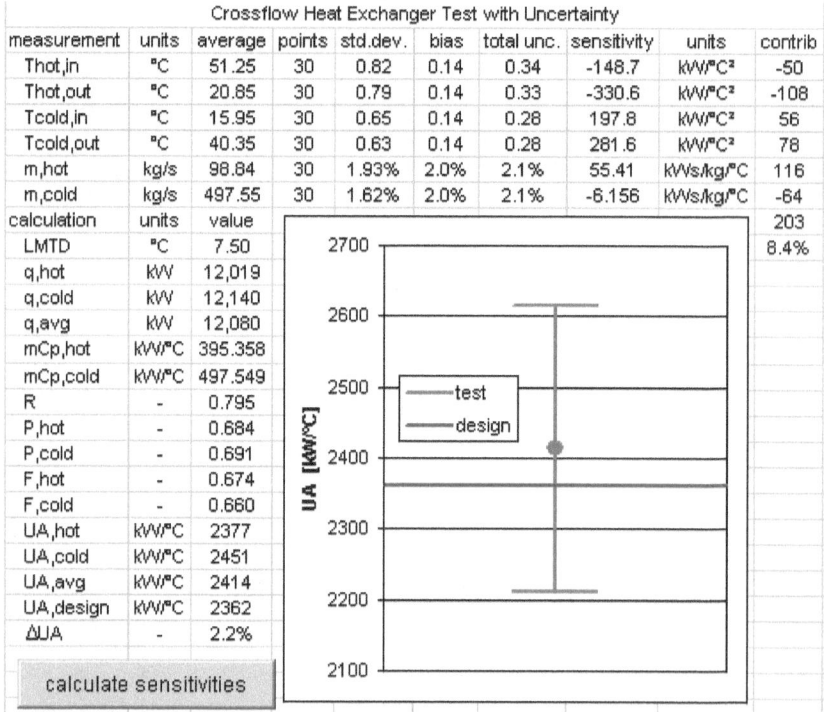

The temporal component of the uncertainty is assumed to be random and proportional to the standard deviation of the average. This is multiplied by Student's-T value at 95% to account for the number of readings. The time-independent component is often called *bias* or *systematic* uncertainty. This represents the accuracy of the instrument and its calibration.

The systematic uncertainty of the temperature measurements is listed as 0.26/1.8°C, which is typical for precision RTDs[1] with NIST-traceable pedigree[2]. The uncertainty of the flow measurement is listed as 2%. If someone tells you they have a flow meter that's more accurate than this, don't believe them. It is evident from the following equation that the measurements cannot possibly be any more accurate than the systematic uncertainty of the device.

$$u_i = \sqrt{b_i^2 + \left(t_{95} \frac{s_X}{\sqrt{N}}\right)^2} \qquad (3.9)$$

Here u_i is the uncertainty, b is the bias (or systematic uncertainty), s_X is the standard deviation of the measured values, t_{95} is Student's-T value at the 95% confidence level[3], and N is the number of points. The sensitivities, θ_I, are partial derivatives of the result, UA_{AVG}, with respect to each measurement. A macro is provided that perturbs each value and calculates the sensitivity, which is multiplied by u_{TOTAL} to obtain each uncertainty contribution. The total uncertainty is the root-sum-square of the components, in this case 203 kW/°C or 8.4% of the average 2414 kW/°C.[4]

$$u_{TOTAL} = \sqrt{\sum_{i=1}^{m} (\theta_i u_i)^2} \qquad (3.10)$$

The average test UA is 2.2% above the design value of 2362 kW/°C, but the test uncertainty is much larger than this at ±8.4%, as shown by the green bars in the figure, so that it is not possible to say that the heat exchanger is performing better than design. This result would, however, constitute adequate proof that the heat exchanger has *passed* the performance test.

---

[1] Resistance Temperature Detector of the 4-wire Platinum element variety.
[2] The calibration process meets the requirements of and can be traced back to the National Institute of Standards and Technology.
[3] Use TINV(0.05,count) to calculate this value in Excel®.
[4] More details on uncertainty calculations may be found in the standard reference for Test Uncertainty, ASME PTC 19.1-2013. I must warn you that this document is very difficult to understand, even when you're already quite familiar with the material.

Chapter 4. Water-Cooled Steam Surface Condenser

A water-cooled steam surface condenser (WCC) is basically a box with a bunch of tubes running through it. These range in size from a couch to a motel. Practical considerations put the condensing steam on the outside of the tubes and the cooling water on the inside. Condensing heat transfer coefficients (on the vapor side) are quite high and don't require large velocities; whereas, convective heat transfer coefficients (on the liquid side) very much depend on velocity to achieve higher values. A typical condenser waterbox and tubesheet is depicted below:

Performance Prediction

The two key references for WCCs are: ASME PTC-12.2 (2010) and Heat Exchange Institute Standards for Steam Surface Condensers 11th Ed. (2012). These two documents are quite different in approach and content, the former is more academic and the latter is more practical. PTC-12.2 presents dimensionless correlations involving Reynolds and Nusselt numbers and the HEI document contains lots of interesting tables and curves.

The primary concern of these documents and various formulas therein is to obtain a value for the overall heat transfer coefficient, U. In practice, there are only three possibilities (arranged in order of decreasing likelihood): 1) you will buy a condenser from somebody who makes them, 2) you will work for somebody who makes them, 3) you will start your own company to make condensers.

If you are buying a condenser, the manufacturer will calculate U using their own secret proprietary formula that the founder (who died long ago) came up with. If you work for that company, you'll walk past the founder's picture on the way to your desk, where you'll use that secret formula to design condensers. If you're starting your own company, you'll average all of the formulas, hoping that at least some of them work, add a 2% margin just to be safe, worry, add another 2% margin, worry some more, and add another 1% margin. Over the years you'll develop your own formula to leave behind along with a picture to inspire future employees.

Since, in the vast majority of cases, you will be given a set of condenser performance curves, we will begin there. Shown below is a typical set of such curves for a large power plant condenser:

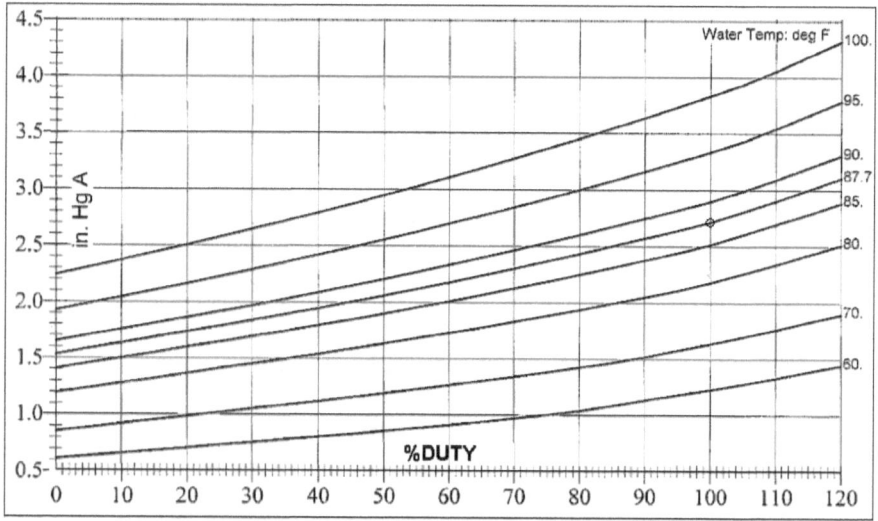

These most often have heat load or percent duty along the X-axis and absolute pressure in inches of mercury (or kPa) on the Y-axis. The curves are at different values of inlet cooling water temperature. There will be a curve that runs through the design point, in this case, 87.7°F, indicated by the little circle.

> *These curves are often provided in mixed units. The modern engineer should be comfortable working with many systems of units. The impression is often given in academia that the formal SI system is used consistently throughout the world, but this is not the case. My advice is to get comfortable with it and don't fuss. There are way too many people fussing already.*

These curves show what the customer is paying for (viz., pressure), but are not the most efficient representation of the information. The upward curvature of the lines in the preceding graph are entirely due to the relationship between the saturation pressure and temperature of steam and are not an artifact of the heat transfer. If, instead of plotting pressure on the Y-axis, we plot saturation temperature, the curvature is completely eliminated, as illustrated in this next figure:

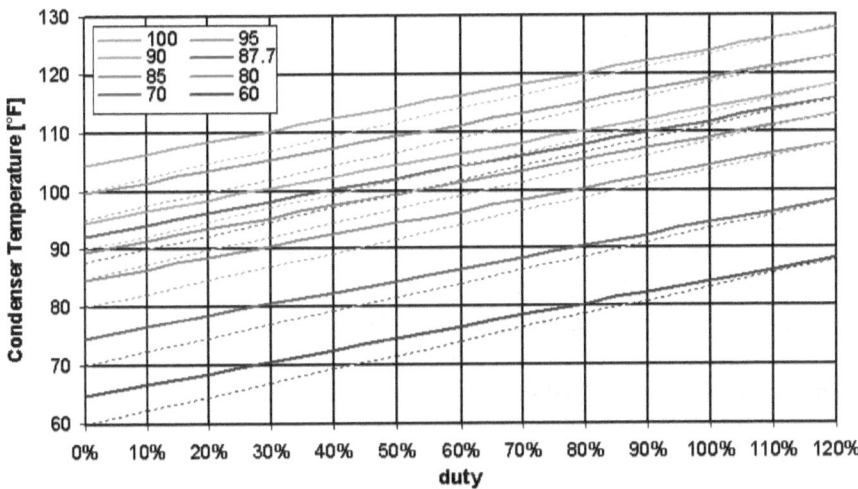

Ignore the dotted lines for the moment. The curves from the previous graph have been reduced to a series of parallel straight lines. While the previous curves could be represented by a multi-variable regression with six to ten terms, this same information as drawn above can be represented by a simple equation with only three terms:

$$T_{SAT} = \alpha + \beta\, duty + \gamma T_W \qquad (4.1)$$

This is to be expected, considering these curves were derived from the relationship, with the product, UA, remaining essentially constant:

$$Q = UA \cdot LMTD \qquad (4.2)$$

The dotted lines in the graph above are an interesting detail. Some manufacturers provide curves with a gradual uplift on the left side, while others have an abrupt change in slope at low duty. In either case, these represent the same thing. Clearly, if there were no heat transfer (i.e., 0% duty) and there were no non-condensable gases, each line should intersect the Y-axis at the point corresponding to the inlet water temperature, hence, the dotted lines.

The dotted lines are the heat transfer that would be achieved in a *perfect world*. Since we don't live on that planet, we use the solid lines. No one will guarantee the performance of a condenser at this low duty. Manufacturers only provide curves that go down this far because customers fuss if they don't. Each manufacturer has their own way of "adjusting" curves at low duty (or very low inlet water temperatures). No one cares about condenser pressures under these conditions, at least from a performance perspective. This is one of the reasons why plants are designed with more than one circulating water pump. If the pressure gets too low in the winter, resulting in sonic velocity at the exit of the steam turbine, the operator simply turns off one of the pumps. The performance of the steam turbine improves plus there is less auxiliary load: a win-win.

There's a spreadsheet (water_cooled_condenser.xls) in the on-line archive that contains these figures plus macros that illustrate WCC calculations.

Performance Evaluation

Accurately testing a water-cooled steam surface condenser is a lot more complicated than it might seem. Below are some of the difficult questions you must answer in order to do so:

Q1: What is the actual steam flow during the test (not the amount shown on the twenty-year-old heat balance that wasn't accurate on the day it was printed)?

Q2: How are you going to account for all of the other miscellaneous (un-instrumented) flows into and out of the condenser during the test as well as the change in the level of the oddly shaped hotwell?

Q3: What is the quality of the steam as it hits the top of the tube bundle (generally assumed to be at the UEEP or used energy end point)? You're going to need everything from generator loss curves to annulus area, last stage blade length, and an exhaust loss curve, which you'll have to assume is accurate.[5]

Q4: How are you going to figure out how much water is flowing through the condenser?

Q4a: Are you planning to use an acoustic flow meter? Then don't bother testing the condenser. Just scribble down some numbers from the control room on a napkin and come up with an estimate over a few beers, because that's all the accuracy you'll end up with.

Q4b: Are you planning to use the pump curves? Better to rent an acoustic flow meter. Those pump curves were made from a plastic 1:16 scale model and never verified on a prototype.

---

[5] In order to calculate the UEEP you start with the generator output and power factor, obtained from a precision power meter, and add the losses, which are usually of adequate precision. Then you subtract the exhaust loss to get the ELEP, including any adjustment, such as the moisture correction. You use this to build an expansion line along with the previous extractions, which must be dry steam. The whole process is iterative. There's a spreadsheet in the *Thermodynamics* package that does this for you.

Q4c: You're going to need access to a straight run of pipe where you can perform a pitot traverse. There aren't many laboratories that can calibrate your pitot. There's Alden, TVA Norris, Iowa Institute of Hydraulic Research, and the U. S. Naval facility. It's unlikely the latter will be responsive to inquiries, so don't bother asking.

Q4d: Your only other practical option for accurate condenser flow measurement is dye dilution. There are a *few* people who can perform an accurate dye dilution test. This technique is replete with subtleties and surprisingly easy to mess up. Look for experience and you'll be pleased with the results. If you don't have good pitot access, dye dilution is your only option.

Q5: What pressure(s) are you going to use? The pressure varies all over above the tube bank up into the hood all the way to the annulus of the turbine. Are you going to use the static or the stagnation pressure?

Q5a: You might try to measure the stagnation pressure if the velocity doesn't rip the end off your probe or bend the shaft so you can't get it out when the test is over.

Q5b: Are you going to measure the static pressure with a basket? There are several types. The one shown below is omni-directional. Some aren't. I've seen many of the directional ones installed backwards. Granted, it may not be obvious which is the front and back face on these and I've even heard arguments that they work better backwards.

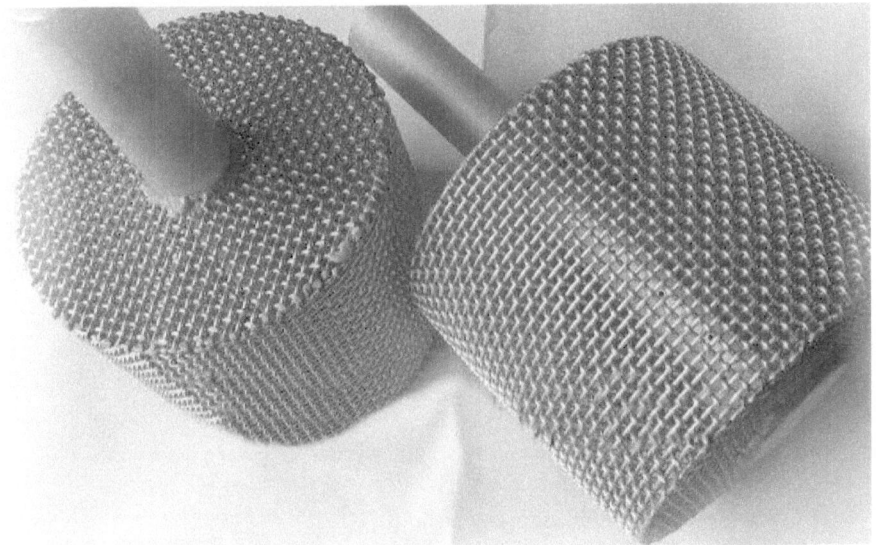

Q5c: Where are you going to measure the pressure? Underneath a flange on the side of the hood where it will be nicely protected but also inside a separation zone, such as the one depicted here behind an airfoil?

I trust you know that the steam exiting the turbine can reach sonic velocity, even faster than that represented by the flow over the airfoil in the following picture. The separation zone shown in this figure and the distorted pressure field is no exaggeration for what can occur in the region where you're trying to measure pressure. You will get as many opinions as to where best locate your pressure sensors as there are people with an interest in the outcome of the test.

The ideal for such a test would be to measure the temperature of the condensate dripping off the top row of tubes. Sadly, this isn't possible. The next best thing would be to measure the temperature of the condensate running off the lowest row of tubes that aren't flooded. I'm sorry to say that I've never had this measurement either.

> The truth is that I'm suspicious of any and all condenser pressure measurements and am skeptical of any that deviate very far from that indicated by the hotwell temperature. When there is a lot of inventory and a large hotwell sub-cooling, this number is useless and I'm skeptical of the whole lot. I only feel good about a test when the pressure lines up fairly well with the hotwell temperature. It's like the testimony of two witnesses, neither one of which is quite believable by themselves, but together are just barely adequate.

Once you have decided where to collect the data and with what instruments and have agreed upon a way to correct for the inflows, outflows, and hotwell inventory, you'll need at least three hours of steady operation. It is also essential to maintain adequate removal of non-condensables during the test. If there is excessive in-leakage, there is no point testing the condenser. You must fix that first.

The following table summarizes the results of an actual test:

	A	B	C	D	E	F	G	H	I
1			CONDENSER THERMAL PERFORMANCE TEST						
2	GEOMETRY	Units	Symbol	Design	Test 1	Test 2	Test 3	Avg.	Notes
3	tubes	-		20,500	20,500	20,500	20,500	20,500	
4	surface area	ft²	A	975,000	975,000	975,000	975,000	975,000	
5	outer diameter	in	Do	1.000	1.000	1.000	1.000	1.000	1"
6	wall thickness	in	wt	0.028	0.028	0.028	0.028	0.028	22g
7	inside diamer	in	Di	0.944	0.944	0.944	0.944	0.944	
8	TEST INPUTS	Units		Design	Test 1	Test 2	Test 3	Avg.	
9	steam pressure	in HgA	Ps	2.50	2.63	2.62	2.67	2.64	meas.
10	water flow rate	gpm	W	335,000	338,147	338,389	338,087	338,208	pitot4xtr
11	water inlet temp.	°F	T1	90.0	90.8	90.8	91.3	91.0	RTDx2
12	water outlet temp.	°F	T2	108.0	109.4	109.3	109.9	109.5	RTDx4
13	PROPERTIES								
14	density	lbm/ft³	ρ	62.00	61.98	61.98	61.98	61.98	
15	specific heat	BTU/lbm/°F	Cp	0.9981	0.9981	0.9981	0.9981	0.9981	
16	CALCULATIONS	Units		Design	Test 1	Test 2	Test 3	Avg.	
17	duty	MBTU/hr	Q	2993	3121	3106	3120	3121	mCpΔT
18	steam temp.	°F	Ts	108.6	110.4	110.3	110.9	110.5	Tsat
19	LMTD	°F	ΔT	5.3	6.3	6.3	6.3	6.3	
20	U	BTU/hr/ft²/°F	U	576.0	510.2	508.4	509.6	509.4	Q/A/ΔT
21	RESULTS	Units		Design	Test 1	Test 2	Test 3	Avg.	
22	Cleanliness	-	Cf	90.0%	88.6%	88.3%	88.5%	88.4%	U/Udes

The spreadsheet (WCC_test.xls) is contained in the archive.

Chapter 5. Air-Cooled Steam Surface Condenser

An air-cooled steam surface condenser (ACC) is basically a car radiator the size of a Walmart on stilts, such as the one pictured below. They're quite expensive to buy and operate. Their chief advantage is that they don't consume water like an evaporative cooling tower. An ACC is like a WCC combined with a dry cooling tower only with fewer moving parts to break—a good choice in arid regions.

Performance Prediction

The performance curves for ACCs are very similar to the ones for WCCs with the addition of fan speed as a variable. The condensation and separating material (e.g., tube wall) conductances are much higher than the convective heat transfer coefficient on the air side, making this the dominating factor. This is even more pronounced than for the water on the inside of the tubes in a WCC. Because air has a much lower thermal conductivity and specific heat than water, it takes a lot more of it and a lot more surface area to carry away the heat of condensation.

Water temperature correction factors (e.g., HEI's) are of some value with a WCC, but empirical air properties are more important with an ACC. The following figure is typical of such curves.

ACCs fan power consumption can be considerable so that variable-speed motors and/or drives are often used to manage auxiliary load. It is, therefore, necessary to quantify performance in terms of both duty and fan speed. This relationship varies with many factors and must be supplied by the ACC manufacturer. The only alternative would be to establish it through a formal testing process using ASME PTC-30.1 (2007), "Air-Cooled Steam Condensers."

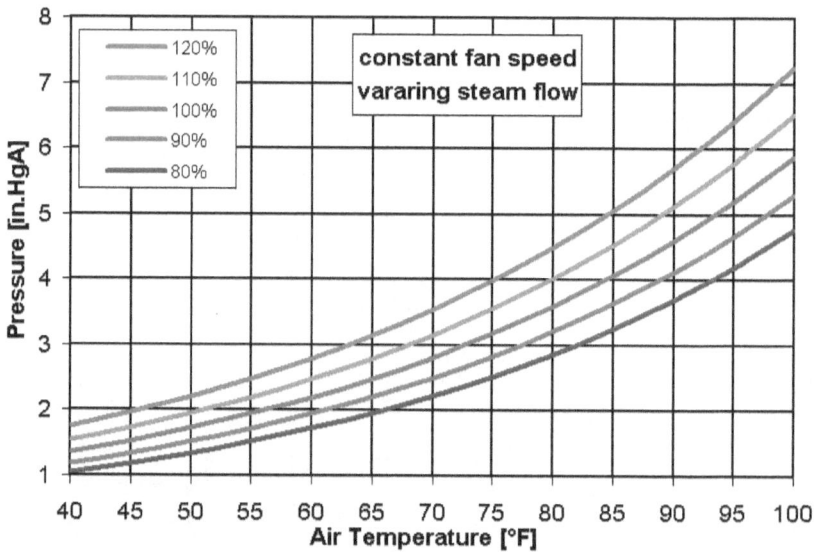

This next figure shows the impact of fan speed at constant steam flow:

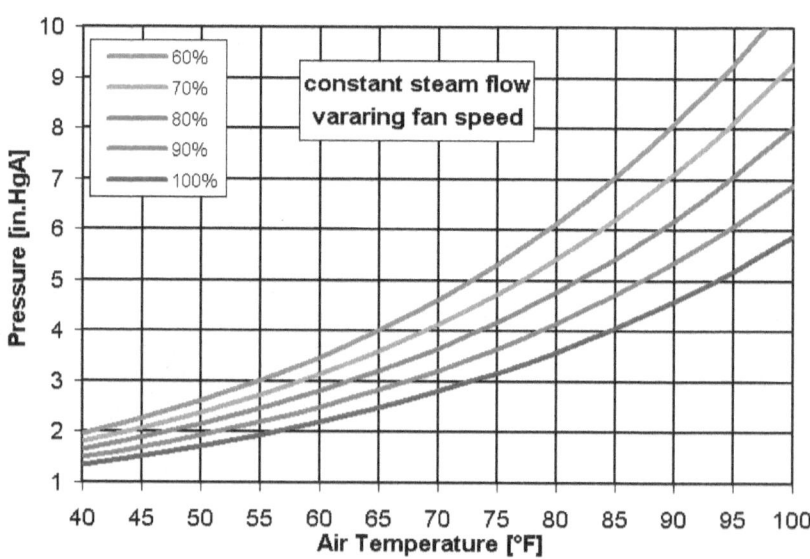

A subtle difference between WCCs and ACCs is that typically the performance of the former is given in terms of percent duty or heat load; whereas, the performance of the latter is most often given in terms of steam flow. The difference being a result of either the entering quality and/or the slight change in enthalpy with the vertical axis, as the corresponding saturation temperature changes.

The preceding two curve sets were reproduced here using multi-variable regression. The manufacturer generated these using empirical correlations for heat transfer coefficients, which means that these can also be inferred from the curves and then a regression performed on the result. Once the product of the effective surface area, A, and overall conductance, U, have been "extracted" from the curves, the same may be used to regenerate the curves from $Q=UA*LMTD$. The spreadsheet contains macros that perform the calculation both ways so that you can compare the two.

Multi-variable regression on ACC curves is a little more challenging than WCC curves, but far from overwhelming. First, get my digitizing tool if you don't already have it. If the curves are in color, reduce the colors to Window's 16 before copying the image onto the clipboard so you can take advantage of the automatic feature (Ctl-A per selected color or Ctl-Alt-A for all 16 colors).

If the image is already on the clipboard before you launch Digitize, it will load it for you and you won't have to open a file. I usually get the images onto the clipboard using the Acrobat snapshot tool, since the graphs usually come in a PDF.

I always define 4 control points. If the R^2 isn't close to 1.00, then one or more of your points is bad. Go back and check them all.

When you exit Digitize, the results are automatically copied onto the clipboard. Just paste them into Excel. You can save it in a file if you want to, but you don't have to. The first two columns are the pixels and the third and fourth are x and y. After you arrange all of the data, use my curve-fitting tool to perform the regression. It will copy the resulting VB macro onto the clipboard for you to paste into Excel.

There's a spreadsheet (air_cooled_condenser.xls) in the on-line archive that contains these figures plus macros that illustrate ACC calculations.

Performance Evaluation

Testing an ACC is in some ways less complicated than testing a WCC—but that's not to say that it's cheaper. First of all, there are lots of places to measure the steam pressure and there are frequently already fittings where you can readily attach instruments. You will have the same problem determining the quality of the entering steam (or the UEEP), but it's often easier to quantify the flow rate, because the condensate is all coming out of a pipe at the bottom.

Don't even try measuring the air flow through an ACC. There are holes and gaps all over it and you'll never account for all of them. Just calculate it from the heat load. You'll have enough trouble getting a velocity-weighted average exiting air temperature. On a big ACC it costs way too much to traverse all of the fans. Pick a geometric pattern (center, edge, lateral, corner, etc.), traverse several fans, and apply the results across the rest based on location. It's just like traversing the fans atop a cooling tower and you use the same equipment.

The biggest problem with ACCs is sub-cooling. You neither pay for nor care about sub-cooling. The standpipe will provide more than adequate net positive suction head (NPSH) for the condensate booster pump(s). Sub-cooling means that a bunch of the expensive heat exchange surface that you paid for is flooded, which is about as useful as having a finished basement full of water.

During an ACC performance test, the manufacturer's representative will make a really big deal about how wonderful all the water in your basement is and go to great lengths to explain how they calculated the amount down to the last drop. They must give this spiel. Just smile and listen.

Here's the problem with sub-cooling... If you test when it's hot and you still have sub-cooling, then your ACC is more than adequate. If you test when it's cold and you have sub-cooling, you don't know that you'll have adequate capacity when the weather gets hot. The manufacturer will try to tell you that X degrees of sub-cooling in the winter will translate into Y inches of mercury less pressure in the summer. The trouble is, they never gave you that graph and they can't provide it to you now, because it's just too complicated to draw.

Of course, the test conditions aren't on any of the graphs that you got. Oh, yeah, there's also the fact that half of the fans were off during the test and they never gave you that graph either. They could give you a graph with one-third of the fans turned off, but their software won't let them turn half of the fans off.

Welcome to the world of performance testing!

An accurate velocity-weighted exit air temperature is important to an ACC performance test for two reasons: 1) it is not likely that you will be testing at the guarantee conditions, so you will need to correct the performance back to the base reference conditions, and 2) you will be interested in the fan power consumption, as this is the second most important aspect of the performance.

The following table summarizes an actual fan test and includes all of the calculations:

	A	B	C	D	E	F	G	H	I	J	K
1	ACC Fan Test (two traverses at right angles)										
2	fan dia.	ft	34.4								
3	hub dia.	ft	3.44								
4	net area	ft²	921.2								
5	baro	psia	14.77								
6	pitot coef.	-	835.5	includes calibration and units							
7	6 point along each radius										
8	pnt	X	Y	temp	p	ΔP	angle	Vtang	Vnorm	Area	ρVA
9	-	ft.	ft.	°F	lb/ft³	in.H2O	deg°	fpm	fpm	ft²	lb/min
10	1	-11.6	11.6	86.5	0.073	0.48	32	2142	1817	38.4	5091
11	2	-10.5	10.5	88.5	0.073	0.46	31	2101	1801	38.4	5028
12	3	-9.3	9.3	88.5	0.073	0.52	11	2234	2193	38.4	6122
13	4	-7.9	7.9	89.5	0.073	0.36	18	1860	1769	38.4	4931
14	5	-6.1	6.1	89.5	0.073	0.07	26	820	737	38.4	2055
15	6	-3.2	3.2	85.0	0.073	0.00	0	49	49	38.4	137
16	7	3.2	-3.2	85.0	0.073	0.00	0	39	39	38.4	110
17	8	6.1	-6.1	85.0	0.073	0.00	0	202	202	38.4	569
18	9	7.9	-7.9	88.0	0.073	0.25	23	1548	1425	38.4	3982
19	10	9.3	-9.3	88.0	0.073	0.45	13	2077	2024	38.4	5656
20	11	10.5	-10.5	87.5	0.073	0.57	10	2337	2301	38.4	6436
21	12	-11.6	11.6	87.0	0.073	0.40	11	1957	1921	38.4	5377
22	pnt	X	Y	temp	p	ΔP	angle	Vtang	Vnorm	Area	ρVA
23	-	ft.	ft.	°F	lb/ft³	in.H2O	deg	fpm	fpm	ft²	lb/min
24	13	11.6	11.6	87.0	0.073	0.35	25	1830	1659	38.4	4644
25	14	10.5	10.5	88.5	0.073	0.37	14	1884	1828	38.4	5105
26	15	9.3	9.3	89.0	0.073	0.38	12	1911	1869	38.4	5213
27	16	7.9	7.9	90.0	0.073	0.32	19	1755	1659	38.4	4620
28	17	6.1	6.1	90.0	0.073	0.15	24	1201	1098	38.4	3056
29	18	3.2	3.2	85.0	0.073	0.00	0	44	44	38.4	123
30	19	-3.2	-3.2	85.0	0.073	0.00	0	31	31	38.4	87
31	20	-6.1	-6.1	85.0	0.073	0.00	0	195	195	38.4	549
32	21	-7.9	-7.9	92.0	0.072	0.26	23	1585	1459	38.4	4047
33	22	-9.3	-9.3	91.5	0.072	0.47	12	2130	2083	38.4	5784
34	23	-10.5	-10.5	90.0	0.073	0.57	16	2342	2251	38.4	6268
35	24	11.6	11.6	88.0	0.073	0.44	40	2054	1573	38.4	4397
36	*Note: The air doesn't flow straight up. You must							88.7			89,383
37	measure the angle and multiply by cos(θ).										

The points of a fan traverse must correspond to equal areas. This is described in several ASME and CTI test codes. The relationships are as follows. The net flow area, excluding the hub, is given by:

$$A = \frac{\pi}{4}\left(D_{FAN}^2 - D_{HUB}^2\right) \quad (5.1)$$

Each of the *n* equal annular areas corresponding to the measurement points is given by:

$$\frac{A}{n} = \pi\left(R_{i+1}^2 - R_i^2\right) \qquad (5.2)$$

The first (inner most) and last (outer most) radii correspond to the hub and fan, respectively:

$$R_1 = \frac{d}{2} \qquad R_N = \frac{D}{2} \qquad (5.3)$$

The measurement is taken at the center of each annulus. The outer radius of each annulus is given by:

$$R_O = \sqrt{\frac{i}{n} D_{FAN}^2 + \frac{n-i}{n} D_{HUB}^2} \qquad (5.4)$$

This test consisted of two perpendicular traverses with 12 points along each diameter (6 per radius), spaced to be representative of equal areas. Note that the air does not flow vertically upward. You must measure the angle at each point and then multiply by the $\cos(\theta)$. The total mass flow rate and mass weighted average temperature are easily calculated with functions built into Excel®. The measurement points are illustrated in this next figure:

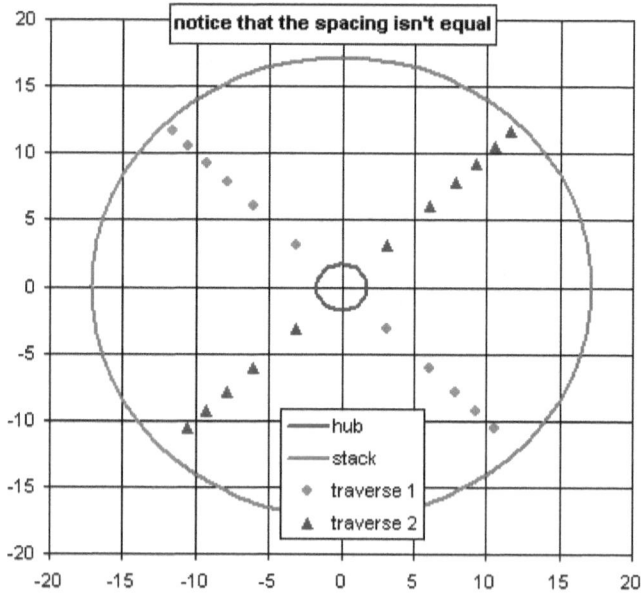

The spreadsheet (ACC_fan_test.xls) is contained in the archive. This is only part of the overall test. There is also the thermal performance on another tab in the same spreadsheet as the performance curves. In this case, two tests were conducted: one in the winter and a second in the summer:

	A	B	C	D	E	F
1			Winter Test			
2	Inputs	Units	Test 1	Test 2	Test 3	Avg.
3	fan spd.	%	75.0%	75.0%	75.0%	75.0%
4	stm. flow	%	102.1%	102.4%	102.6%	102.4%
5	amb. air	°F	48.2	49.5	51.1	49.6
6	stm. pres.	in.HgA	2.19	2.27	2.39	2.28
7	stm. temp.	°F	104.1	105.3	107.1	105.5
8	Calculations	Units	Test 1	Test 2	Test 3	Avg.
9	exp. pres.	in.HgA	2.22	2.31	2.41	2.31
10	exp. pres.	Δ%	-1.3%	-1.5%	-0.9%	-1.2%
11	exp. temp.	°F	104.6	105.9	107.4	105.9
12	exp. temp.	Δ°F	-0.4	-0.5	-0.3	-0.4
13	Corrections	Units	Test 1	Test 2	Test 3	Avg.
14	at design	in.HgA	1.70	1.70	1.71	1.71
15	guarantee	in.HgA	1.73	1.73	1.73	1.73
16	as-tested	Δ%	-1.3%	-1.5%	-0.9%	-1.2%

This ACC passed the winter test by approximately 1.2%, in spite of the fans running at 70% speed and condensing 2.4% additional flow. The summer test was not so impressive:

	A	B	C	D	E	F
18			Summer Test			
19	Inputs	Units	Test 1	Test 2	Test 3	Avg.
20	fan spd.	%	100.0%	100.0%	100.0%	100.0%
21	stm. flow	%	98.0%	97.9%	97.5%	97.8%
22	amb. air	°F	97.3	98.0	98.7	99.3
23	stm. pres.	in.HgA	5.49	5.57	5.66	5.57
24	stm. temp.	°F	137.3	137.9	138.5	137.9
25	Calculations	Units	Test 1	Test 2	Test 3	Avg.
26	exp. pres.	in.HgA	5.40	5.48	5.56	5.48
27	exp. pres.	Δ%	1.7%	1.6%	1.8%	1.7%
28	exp. temp.	°F	136.6	137.2	137.8	137.2
29	exp. temp.	Δ°F	0.6	0.6	0.7	0.7
30	Corrections	Units	Test 1	Test 2	Test 3	Avg.
31	at design	in.HgA	1.76	1.76	1.76	1.76
32	guarantee	in.HgA	1.73	1.73	1.73	1.73
33	as-tested	Δ%	1.7%	1.6%	1.8%	1.7%

The same ACC failed a second performance test ten months later by 1.7% with the fans at 100% and condensing only 97.8% of the design steam flow. There are more variables impacting the performance of ACCs than WCCs and this should not be overlooked when considering a performance testing plan.

Chapter 6. Simple Shell-and-Tube Heat Exchanger

Condensers were introduced before simple shell-and-tube heat exchangers, as their analysis is often less complicated. For one thing, the LMTD and NTU-ε methods are the same for condensers because C_P on the condensing side is infinite. Because the temperature on one side of a condenser is constant, geometry isn't an issue either.

The simplest type of shell-and-tube heat exchanger to analyze is one that has sufficient baffles to approximate a counter-current double-pipe geometry, as illustrated schematically in this next figure:

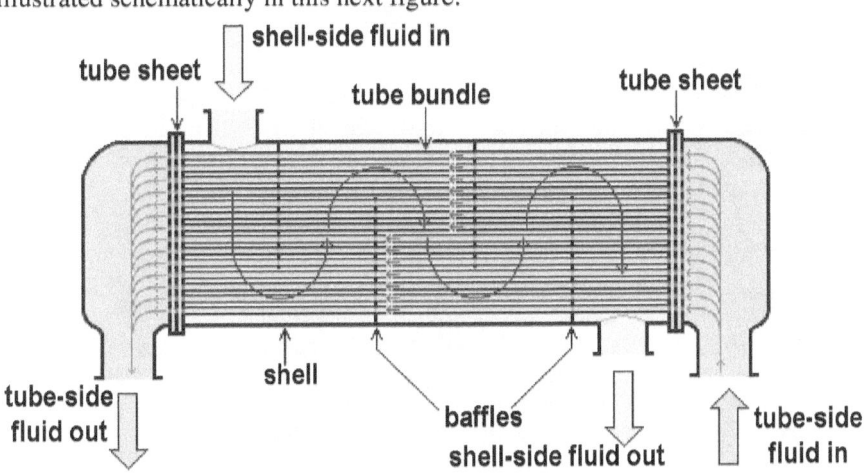

27

Performance Prediction

The heat transfer coefficient inside the tubes is fairly simple to calculate and there are several equations to choose from. No one should ever design a heat exchanger where the flow inside the tubes isn't turbulent, so there's no point developing a correlation for anything but turbulent flow. The classic equation was developed by Dittus&Boelter[6]:

$$Nu = 0.023 \, \text{Re}^{\frac{4}{5}} \, \text{Pr}^n \quad (6.1)$$

where n=0.4 for heating and n=0.3 for cooling. A more recent correlation has been developed by Sieder&Tate[7]:

$$Nu = 0.027 \, \text{Re}^{\frac{4}{5}} \, \text{Pr}^{\frac{1}{3}} \left(\frac{\mu_{BULK}}{\mu_{SURFACE}} \right)^{0.14} \quad (6.2)$$

Yet another correlation has been developed by Rabas&Cane[8]:

$$Nu = 0.0158 \, \text{Re}^{0.835} \, \text{Pr}^{0.462} \quad (6.3)$$

Resistance of the tube wall can be found in any heat transfer text:

$$R_W = \frac{D_O}{2k_W} \ln\left(\frac{D_O}{D_I}\right) \quad (6.4)$$

There are numerous correlations for the shell side heat transfer coefficients. You'll have to find one that is appropriate for your particular application. These vary with the fluids, baffles, and tube spacing. Fouling is often used as a *fudge* factor. The overall heat transfer coefficient is found by summing the resistances:

$$\frac{1}{U} = \left(\frac{D_O}{D_I}\right)(R_{FI} + R_T) + R_M + R_S + R_{FO} \quad (6.5)$$

Here R_{FI} and R_{FO} are the inside and outside fouling resistance, respectively. R_T, R_W, and R_S are the tube side, tube material, and shell side resistances, respectively. The D_O/D_I term accounts for the fact that the inner and outer surface areas of a tube aren't the same per unit length.

---

[6] Dittus, P. W. and L. M. Boelter, L.M., University of California Publications in Engineering, Vol. 1, No. 13, pp. 443-461 1930 (reprinted in *International Communications in Heat and Mass Transfer*, Vol. 12, pp. 3-22, 1985).

[7] Sieder, E. N. and G. E. Tate, "Heat Transfer and Pressure Drop of Liquids in Tubes," Industrial Engineering Chemistry, Vol. 28, p. 1429, 1936.

[8] Rabas, T. J., and D. Cane, "An Update of Intube Forced Convection Heat Transfer Coefficients of Water," *Desalinization*, Vol. 44, pp. 109–119, 1983.

There are also several methods for estimating the shell side heat transfer coefficient. One of the earliest methods was developed by Kern[9], based on industrial heat exchangers and is quite similar to the Sieder-Tate.

$$Nu = 0.36 \, Re^{0.55} \, Pr^{\frac{1}{3}} \left(\frac{\mu_{BULK}}{\mu_{SURFACE}} \right)^{0.14} \quad (6.6)$$

This correlation could also be cast in the same form as Dittus-Boelter for convenience:

$$Nu = 0.36 \, Re^{0.55} \, Pr^{n} \quad (6.7)$$

The length in both the Nusselt and Reynolds number is an equivalent diameter (D_E), that takes into account the tube spacing, pitch (P), and bundle alignment. This first equation (6.7) is for a square tube arrangement:

$$D_E = \frac{4P^2}{\pi D_O} - D_O \quad (6.8)$$

and this second (6.8) is for a triangular arrangement:

$$D_E = \frac{2\sqrt{3}P^2}{\pi D_O} - D_O \quad (6.9)$$

The shell-side velocity, V_s, is given by:

$$V_S = \frac{\dot{m}_S}{\rho N_R (P - D_O) \left(\dfrac{L_T}{N_B} \right)} \quad (6.10)$$

where N_R is the number of tubes per row (across the flow), L_T is the tube length, and N_B is the number of baffles. You may want to adjust the effective shell-side area to better represent the actual tube bundle and baffle arrangement in your heat exchanger.

Performance Evaluation

We will first consider a side-stream oil cooler. As usual, we are only given minimal information by the manufacturer and have been asked to conduct a performance test to determine if it is fouled and how badly. We are given the design operating conditions, from which we will infer the overall heat transfer coefficient and as-new fouling resistance. For this simple arrangement and almost constant properties the LMTD and NTU-ε methods yield essentially the same results.

---

[9] Kern, D. Q., *Process Heat Transfer*, McGraw-Hill, 1950.

The following table shows the geometry, properties, and performance calculations at the design conditions:

Side-Steam Oil Cooler - Design Calculations						
geometry	units	value	dimensions	units	result
tubes	-	16	inside dia.	mm	9.40
tubes/row	-	4	tube clear.	mm	4.45
outside dia.	mm	12.7	baffle spacing	m	0.46
tube wall	mm	1.7	shell eff. dia.	mm	12.8
tube pitch	mm	17.1	calculations	units	result
tube len.	m	3.66	heat transfer	kW	50.8
baffles	-	8	water flow	kg/hr	7,521
operating point	units	value	LMDT	°C	84.3
oil flow	kg/hr	3,402	UA	W/C°	602
oil inlet temp.	°C	115.6	A	m²	2.33
oil exit temp.	°C	90.6	U	W/m²/°C	258
water in tmp.	°C	15.6	tube vel.	m/s	0.99
water out tmp.	°C	21.1	tube Re	-	2,546
properties	units	value	tube Nu	-	40.1
tube cond.	W/m/°C	25.9	tube side h	W/m²/°C	546
oil density	kg/m³	859	tube side R	°C-m²/W	0.0018
oil spec. ht.	kJ/kg/°C	2.15	tube wall h	W/m²/°C	4732
oil viscosity	cp	3.14	tube wall R	°C-m²/W	0.0002
oil cond.	W/m/°C	0.128	shell vel.	m/s	0.26
oil Prandtl	-	52.8	shell Re	-	3,145
water density	kg/m³	998	shell Nu	-	68
water sp. ht.	kJ/kg/°C	4.38	shell side h	W/m²/°C	3179
water visco.	cp	1.05	shell side R	°C-m²/W	0.0003
water cond.	W/m/°C	0.597	fouling h	W/m²/°C	1140
water Prandtl	-	7.68	fouling R	°C-m²/W	0.0009

The properties are calculated using interpolation and tables that are also provided in the same spreadsheet. The geometry (number and size of tubes), inlet and outlet temperatures, and oil flow are provided. The heat transfer is calculated from the oil side and the water flow rate from that. The overall heat transfer coefficient is calculated from $U=Q/A/LMTD$, the tube side, wall, and shell side heat transfer coefficients are calculated, leaving only the fouling resistance, R_W.

The four one-hour test periods are summarized in the following table:

Side-Stream Oil Cooler - Test Calculations						
Measurements	units	Test 1	Test 2	Test 3	Test 4	Average
oil flow	kg/hr	3,241	3,241	3,235	3,245	3,240
oil inlet temp.	°C	112.9	113.2	113.0	113.9	113.3
oil exit temp.	°C	89.8	90.5	91.4	91.6	90.8
water in tmp.	°C	14.4	15.5	17.5	15.0	15.6
water out tmp.	°C	18.9	20.1	22.0	19.5	20.1
properties	units	value				
tube cond.	W/m/°C	15.0	16.0	17.0	18.0	16.5
oil density	kg/m³	860.38	860	860	860	860
oil spec. ht.	kJ/kg/°C	2.147	2.15	2.15	2.15	2.15
oil viscosity	cp	3.24	3.21	3.19	3.16	3.20
oil cond.	W/m/°C	0.128	0.128	0.128	0.128	0.128
oil Prandtl	-	54.4	53.9	53.6	53.1	53.8
water density	kg/m³	998.77	999	998	999	999
water sp. ht.	kJ/kg/°C	4.38	4.38	4.38	4.38	4.38
water visco.	cp	1.09	1.06	1.01	1.08	1.06
water cond.	W/m/°C	0.594	0.596	0.599	0.595	0.596
water Prandtl	-	8.05	7.80	7.38	7.92	7.79
calculations	units	result				
heat transfer	kW	44.6	44.0	41.8	43.3	43.4
water flow	kg/hr	8,054	7,873	7,510	8,062	7,875
LMDT	°C	84.4	83.8	82.2	85.2	83.9
UA	W/C°	529	526	508	508	518
U	W/m²/°C	227	225	218	218	222
tube vel.	m/s	0.94	0.94	0.94	0.94	0.94
tube Re	-	2,349	2,372	2,383	2,415	2,380
tube Nu	-	38.0	38.1	38.2	38.5	38.2
tube side h	W/m²/°C	517	519	520	524	520
tube side R	°C-m²/W	0.0019	0.0019	0.0019	0.0019	0.0019
tube wall h	W/m²/°C	4732	4732	4732	4732	4732
tube wall R	°C-m²/W	0.0002	0.0002	0.0002	0.0002	0.0002
shell vel.	m/s	0.28	0.27	0.26	0.28	0.27
shell Re	-	3,231	3,247	3,256	3,283	3,254
shell Nu	-	71	70	68	71	70
shell side h	W/m²/°C	3271	3250	3201	3283	3251
shell side R	°C-m²/W	0.0003	0.0003	0.0003	0.0003	0.0003
fouling H	W/m²/°C	781	758	680	666	721
fouling R	°C-m²/W	0.0013	0.0013	0.0015	0.0015	0.0014
					-14%	59%

As shown at the bottom of the table, the fouling resistance has increased from 0.0009 °C-m²/W to 0.0014 (59%) for a reduction in the overall conductance from 258 W/m²/°C to 222 (14%).

Chapter 7. Feedwater Heater

There are three factors that can greatly complicate heat exchanger analysis. We have already discussed two: varying properties and geometry. The third is zones of phase change within the heat exchanger. Each of these factors is ignored (glossed over, set aside) during the process of separation and integration of the governing differential equation in order to obtain a closed-form solution.

Feedwater heaters are perhaps the most common type of heat exchanger that has all three zones (de-superheating, condensing, and sub-cooling) within the same shell. Heat Recovery Steam Generators (HRSGs) used in combined cycle power plants also have these three zones, as to all conventional boilers, but the three are separated into discrete sections in those designs.

Performance Prediction

In order to explore this type of heat exchanger, it is helpful to consider a graph with heat transfer along the horizontal axis and temperature along the vertical, as in the following figure:

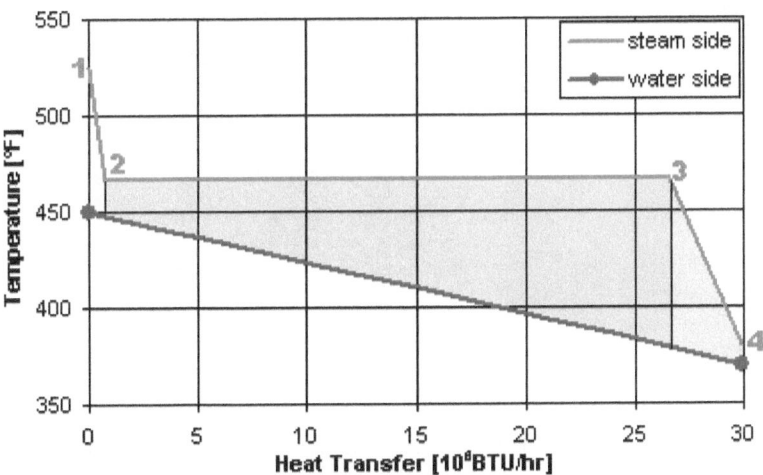

The yellow area represents the de-superheating zone, in which the steam is cooled from a superheated state to saturated vapor. The red process line has a sharp downward slope between points 1 and 2. The green area represents the condensing zone, in which the steam condenses on the outside of the tube bundle. The red process line is flat over this zone because the saturation temperature is constant. The cyan area represents the sub-cooling zone, in which the condensate is further cooled, as it approaches the inlet feedwater temperature. The red process line from point 3 to 4 has a downward slope that is not as steep as the line from point 1 to 2 because the specific heat of the condensate is larger than the specific heat of the vapor and the rate of change of temperature along the process line is given by:

$$\frac{dT}{dq} = \frac{-1}{\dot{m}C_P} \tag{7.1}$$

Neither the F-LMTD nor the P-NTU method can be applied to this heat exchanger as a whole because there are three thermally discontinuous zones set apart by the phase change. Inside the shell there are often partitions plus there is always an interface below which the tube bundle is flooded. However, these methods can be used effectively on each of the three sections separately, as illustrated in the spreadsheet, feed_water_heater_analytical.xls, which is included in the archive.

	A	B	C	D	E	F	G	H	I
1			Feed Water Heater Example						
2	INPUTS	units		CALCULATIONS	units				
3	feed water inlet			feed water inlet					
4	flow	lbm/hr	350,000	enthalpy	BTU/lbm	345.9			
5	pressure	psia	2250	feed water outlet					
6	temperature	°F	370	enthalpy	BTU/lbm	431.4			
7	feed water outlet			extraction					
8	pressure	psia	2200	saturation	°F	467.0			
9	temperature	°F	450	TTD	°F	17.0			
10	extraction			superheat	°F	58.0			
11	pressure	psia	500	enthalpy	BTU/lbm	1249.7			
12	temperature	°F	525	flow	lbm/hr	33,382			
13	drain			drain					
14	temperature	°F	380	DCA	°F	10.0			
15				enthalpy	BTU/lbm	354.0	U	A	area
16	heat trans. zone	Qx10⁶	UA	zone color	units	ΔT	BTU/hr/ft²/°F	ft²	frac
17	de-superhtng	1.5	0.035	yellow	°F	42.4	100	354	23%
18	condensing	25.2	0.537	green	°F	46.9	600	895	58%
19	subcooling	3.2	0.089	cyan	°F	36.0	300	295	19%
20	process line		0.661	calculations		T		1544	
21	point 1	0	525	sat. vapor	BTU/lbm	1204.7			
22	point 2	1.5	467	sat. liquid	BTU/lbm	449.5			
23	point 3	26.7	467	desup. exit temp	°F	446.0			
24	point 4	29.9	380	cond. exit temp.	°F	378.5			

The LMTD in the de-superheating, condensing, and sub-cooling (yellow, green, and cyan) zones is 42.4, 46.9, and 36.0°F, respectively. The heat transfer in these three zones is 1.5, 25.2, and 3.2x10⁶ BTU/hr, respectively. The conductance, $UA=Q/LMTD$, in these three zones is 0.035, 0.537, and 0.089x10⁶ BTU/hr/°F, respectively for a total of 0.661x10⁶ BTU/hr/°F.

Typical overall heat transfer coefficients, U, for the de-superheating, condensing, and sub-cooling zones would be: 100, 600, and 300 BTU/hr/ft²/°F. This would indicate required surface areas of 354, 895, and 295 ft², respectively, for a total of 1544 ft². As there is typically a single tube U-shaped bundle in these heat exchangers, this means that 23% of the tube area at the top near the steam inlet is surrounded by dry steam on the shell side, 58% of the tube area is dripping with condensate, and 19% at the bottom is flooded.

Feedwater heater performance is most often specified in terms of Terminal Temperature Difference (TTD) and Drain Cooler Approach (DCA). By industry convention, TTD, rather than being the actual terminal difference, is the saturation temperature of the steam in the condensing zone minus the feedwater outlet temperature. This value can be zero and even negative if there is sufficient superheat of the extraction steam. TTD is not used in the sense of $Q \neq UA*\Delta T$, so a value less than or equal to zero does not indicate a violation of the Second Law of Thermodynamics. The following figure shows the relative impact of TTD and DCA on the required size of the heat exchanger in this example.

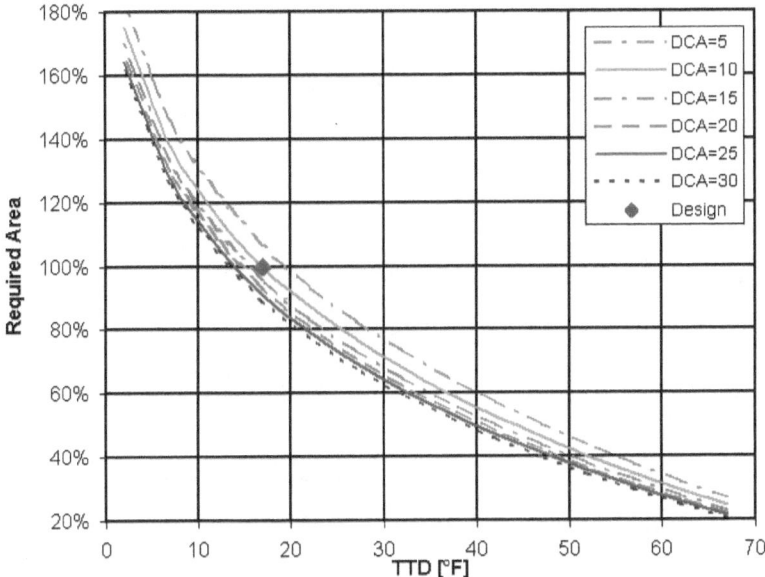

Performance Evaluation

Performance test calculations for a feedwater heater are similar to the crossflow and TEMA-E examples except that the process must be separated into the three zones (de-superheating, condensing, and sub-cooling).

Feed Water Heater Test with Uncertainty									
measurement	units	average	points	std.dev.	bias	total unc.	sensitivity	units	contrib
Tfw,in	°F	371.5	30	0.63	0.14	0.28	-213.021	BTU/hr/°F/°F	-58.74
Tfw,out	°F	451.3	30	0.47	0.14	0.23	19948.34	BTU/hr/°F/°F	4530
Textract	°F	524.8	30	0.58	0.14	0.26	-964.167	BTU/hr/°F/°F	-250.7
Tdrain	°F	380.3	30	0.71	0.14	0.30	-3949.15	BTU/hr/°F/°F	-1191
Pfw,in	psia	2248	30	0.93%	0.75%	0.01	-11.601	BTU/hr/°F/psia	-0.096
Pfw,out	psia	2201	30	0.86%	0.75%	0.01	5.699826	BTU/hr/°F/psia	0.046
Pextract	psia	499	30	0.74%	0.75%	0.8%	-2985.15	BTU/hr/°F/psia	-23.86
flow,fw	lbm/hr	337,213	30	0.98%	2.0%	2.0%	1.966478	BTU/hr/°F/lbm/hr	0.04
calculation	units	value							4691
hfw,in	BTU/lbm	347.5							5.2%
hfw,out	BTU/lbm	432.8							
hextract	BTU/lbm	1249.6							
hg	BTU/lbm	1204.7							
hf	BTU/lbm	449.3							
hdrain	BTU/lbm	354.3							
flow,extract	lbm/hr	32,123							
Qdesuper	BTU/hr	1.4E+06							
Qconds	BTU/hr	2.4E+07							
Qsubcool	BTU/hr	3.1E+06							
Tsat	°F	466.8							
TTD	°F	15.5							
DCA	°F	8.8							
Tfw,desup	°F	447.3							
Tfw,subco	°F	380.0							
LMTD,desup	°F	40.7							
LMTD,cond	°F	45.1							
LMTD,subc	°F	34.1							
UA,desup	BTU/hr/°F	3.5E+04							
UA,conds	BTU/hr/°F	5.4E+05							
UA,subco	BTU/hr/°F	9.0E+04							
UA,total	BTU/hr/°F	6.6E+05							

Chapter 8. Heat Recovery Steam Generator

Heat Recovery Steam Generators (HRSGs) are common in twenty-first century power plants. Gas Turbines (GTs) are fairly efficient; yet continuously belch out large quantities of hot exhaust. Run this hot air into a heat exchanger, make steam, pass that through a turbine attached to a generator, and achieve even higher efficiency. The combination (GT+HRSG) is called a combined cycle power plant (CCPP) and is even more efficient than a coal-fired power plant (CFPP). The following is a typical HRSG: The GT exhaust enters on the bottom left and the stack gas leaves at the top right.

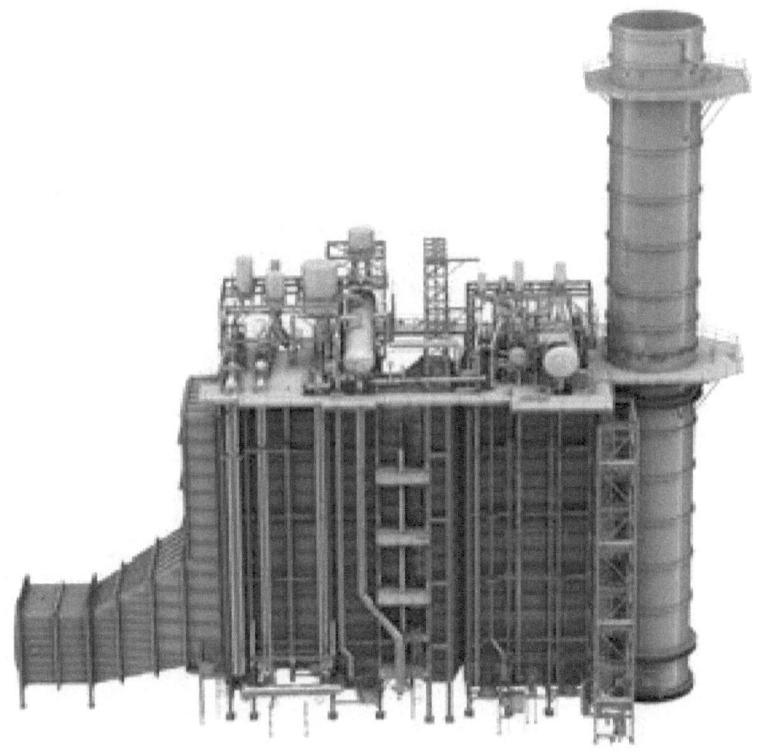

In the current fuel market with cheap natural gas and regulators putting the squeeze on emitters of what are perceived to be greenhouse gases, combined cycle power plants are far more economical than ones burning coal. Of course, some day the gas wells may run out or wars may impact the supply or some celebrity's house may fall into a sink hole near a gas field where there's been cracking so that the price of gas goes through the roof and we may be sorry to have torn down all the coal plants.

While combined cycle plants are thermodynamically efficient, they are by no means cheap. Turbines, heat exchangers, condensers, and cooling towers all cost a hundred million dollars. There are two basic types of HRSGs: single and triple pressure. The former deliver steam to a turbine without reheat or for process (e.g., a paper mill or desalination plant) and the latter to a steam turbine with reheat.

The three pressures in the HRSG are necessitated by the steam turbine design, not the aspects of the heat exchanger. Details of the reheat Rankine cycle are beyond the scope of this book, but suffice it to say that this is to keep the back end of the turbine out of the wet steam zone as much as possible. In a HRSG there are three distinct zones of differing heat exchange: superheating, evaporating, preheating (equivalent of de-sub-cooling or bringing the sub-cooled liquid up to the saturated liquid state).

The three heat exchange processes in a HRSG are accomplished in different types of heat exchangers: superheater, evaporator, and economizer. Most often, these will be three separate devices. Water flows inside tubes and exists only as a single phase in a superheater (vapor) and economizer (liquid), but transitions from liquid to vapor in an evaporator. The GT exhaust is always in a gaseous phase and flows through the shell in all three types. A typical single-pressure design is shown below:

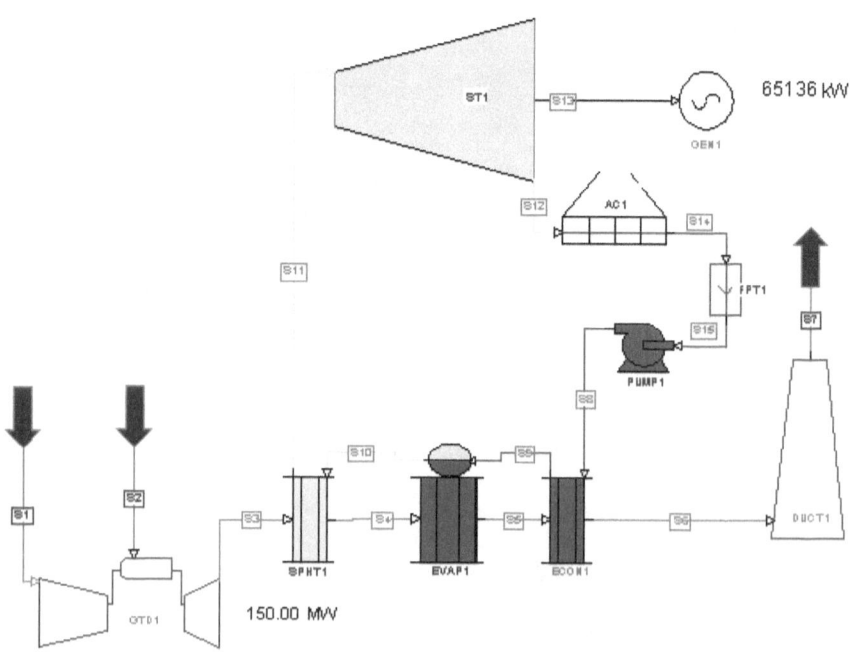

Performance Prediction

Thermal performance of a HRSG is illustrated by what is called a *heat release diagram*, which is a graph showing cumulative heat transferred from the gas to the steam on the horizontal axis and temperature on the vertical axis. The hot gas process line is drawn in red and the steam process line is drawn in blue.

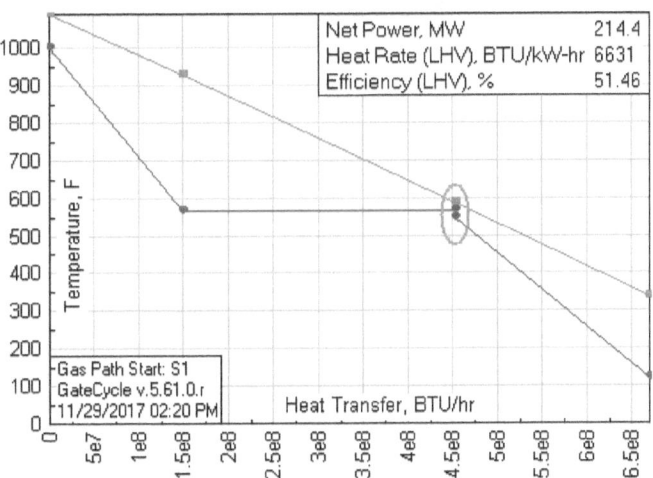

The slope of the hot process line is continuous and almost linear because the specific heat of the GT exhaust gas varies only slightly over this range of temperatures. The steam process line is discontinuous, owing to the fact that the first section is the superheater, the second is the evaporator, and the third is the economizer. This figure is very similar to the one in the preceding chapter on feedwater heaters and for the same reasons.

The area identified by the magenta ellipse is called a *pinch point* and occurs only this once in a single-pressure HRSG. As with the feedwater heater, each section can be analyzed using either the LMTD or P-NTU methods. In either case, the temperature difference at the pinch point is small, resulting in a large required UA. In fact, the performance of HRSGs is dominated by pinch points.

The pinch point shown above will limit the amount of steam that can be generated. A *tighter* (i.e., smaller) pinch will produce more steam produced and a *looser* (i.e., larger) pinch will produce less steam. The amount of steam produced is almost linearly proportional to the pinch, as illustrated in the following figure. The required surface area (or more accurately required UA) is quite a different matter. A tighter pinch would require a much larger UA, which would mean much more expensive heat exchangers.

Designing a cost-effective HRSG is an art and requires juggling more than just surface areas.

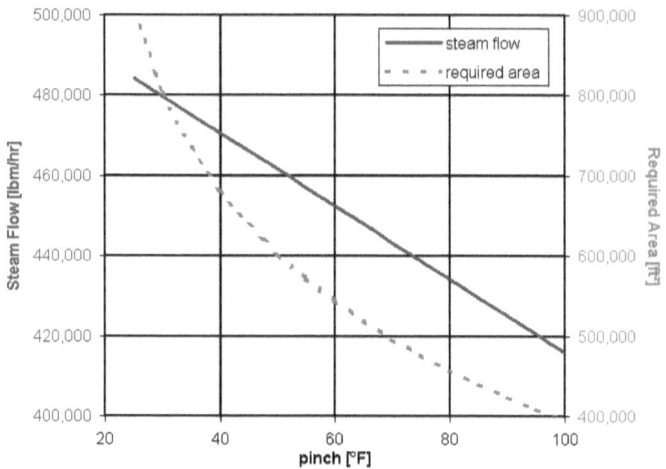

The following is a typical three-pressure HRSG heat release diagram:

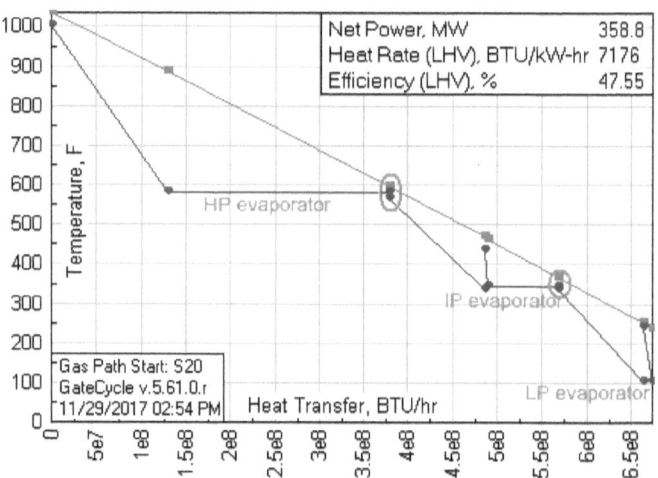

Thermal analysis of the three components that are combined to create a HRSG (superheater, evaporator, and economizer) is the same as for the individual components or a feedwater heater. The arrangement, construction specifics, and properties may be different, but these are all crossflow heat exchangers, as the water and steam flows inside vertical tubes and the exhaust gas flows across the outside of the tubes.

Performance Evaluation

The following is typical of data collected for a HRSG performance test as conducted in accordance with ASME PTC 4.4-2008 (see HRSG_test.xls):

	A	B	C	D	E	F	G	H	I	J	K	L
1		EGW	EGT	STACK	STEAM	FW	EVAP	SPHT	FW	ECON	EVAP	SPHT
2		lb/hr	°F	°F	lb/hr	psia	psia	psia	°F	°F	°F	°F
3	7/11/2017 11:00	3580009	1062.184	347.2219	474780.8	1410.52	1268.351	1204.69	109.8975	554.5939	574.2497	994.1239
4	7/11/2017 11:01	3563845	1062.229	347.3358	475826.9	1413.847	1271.057	1207.279	110.0538	554.7798	574.5233	994.1703
5	7/11/2017 11:02	3556583	1062.518	347.5338	475015.8	1412.153	1269.844	1206.209	110.5083	554.6401	574.4008	996.0175
6	7/11/2017 11:03	3560494	1062.37	347.8043	475418.8	1412.809	1270.061	1206.332	110.2606	554.6644	574.4227	994.2704
7	7/11/2017 11:04	3552742	1062.557	347.6749	474764.9	1410.732	1268.572	1204.929	110.5826	554.5766	574.2722	994.5612
8	7/11/2017 11:05	3564397	1062.12	348.0719	475776.4	1414.592	1271.832	1208.113	109.8982	554.8629	574.6016	995.9677
9	7/11/2017 11:06	3568099	1062.197	348.0178	476385.4	1415.531	1272.409	1208.563	110.021	554.8383	574.6599	993.8743
10	7/11/2017 11:07	3549199	1062.758	347.8878	474066.5	1408.799	1267.053	1203.505	110.8345	554.4473	574.1184	995.289
11	7/11/2017 11:08	3560778	1062.615	348.8424	474816.8	1411.941	1269.751	1206.163	110.5964	554.8085	574.3914	996.7969
12	7/11/2017 11:09	3559480	1062.706	348.1149	475709.6	1413.505	1270.785	1207.022	110.8229	554.6433	574.4958	994.2205
13	7/11/2017 11:10	3558904	1063.007	348.0472	475702.3	1414.393	1271.677	1207.969	110.836	554.7864	574.5859	996.053
14	7/11/2017 11:11	3560813	1062.862	348.0524	475932.4	1414.32	1271.467	1207.685	111.0774	554.7416	574.5647	994.3447
15	7/11/2017 11:12	3552305	1063.119	347.7818	475405	1413.676	1271.136	1207.475	111.5319	554.6536	574.5313	996.6328
16	7/11/2017 11:13	3552606	1062.882	348.3405	474236.8	1409.904	1268.057	1204.523	111.0212	554.6046	574.22	996.4935
17	7/11/2017 11:14	3558228	1062.961	348.1659	474855.1	1411.91	1269.897	1206.096	111.1537	554.7639	574.3859	996.5723
18	7/11/2017 11:15	3552335	1062.948	348.3035	474776	1411.633	1269.467	1205.873	111.1917	554.6356	574.3625	996.5461
19	7/11/2017 11:16	3556490	1063.007	348.0992	475358.4	1412.652	1270.14	1206.432	111.2825	554.7296	574.4307	994.7662
20	7/11/2017 11:17	3550507	1063.192	347.9697	474352.8	1410.104	1268.189	1204.632	111.5227	554.6177	574.2333	996.0248
21	7/11/2017 11:18	3546866	1063.384	348.0148	473955.2	1409.336	1267.656	1204.173	111.8187	554.495	574.1794	997.0692
22	7/11/2017 11:19	3562278	1062.778	348.6291	475532.5	1413.35	1270.735	1207.016	110.8908	554.8507	574.498	994.9742
23	7/11/2017 11:20	3551808	1063.544	347.9311	475004.1	1412.144	1269.843	1206.211	112.1186	554.6127	574.4006	996.0909

EGW and EGT are the industry-standard acronyms for GT exhaust flow and temperature, respectively. The GT exhaust flow cannot be measured, rather it is calculated from an energy balance as part of a simultaneous performance test in accordance with ASME PTC 22-2005.

The steam flow is determined from the feedwater flow, as measured through a precision nozzle, and may be adjusted to account for makeup and/or blowdown. Flow measurement of the feedwater in the liquid state is far more accurate than steam in the vapor state, which is why this measurement point is preferred.

*Flow meters are simply **not** accurate, regardless of what you might have been told. For flow measurements consult ASME PTC 19.5-2004, MFC-3M (2004), or MFC-11M (2003). The only accurate meter is a Coriolis and you will not find one on the feedwater or steam line in a HRSG. That's why there should always be a precision nozzle somewhere on the feedwater line.*

In accordance with the test code, data are divided into one-hour test periods. Averages are calculated using =AVERAGE(INDIRECT(ADDRESS(...))) and defining the appropriate rows and columns on the data tab. The analysis tab is divided into sections: data, properties, and calculations. The input section is:

	A	C	D	E	F	G	H
1			HRSG Performance Test				
4	Test Inerval		Test 1	Test 2	Test 3	Test 4	average
5	begin		11:00:00	12:00:31	12:58:57	14:00:31	
6	end		12:00:31	12:58:57	14:00:31	15:00:00	
7	INPUT DATA	units					
8	GT Exhaust						
9	flow	lbm/hr	3,541,849	3,502,088	3,502,019	3,543,139	3,522,274
10	temperature	°F	1064.0	1067.7	1067.8	1064.0	1065.9
11	Stack						
12	temperature	°F	348.4	349.9	349.9	348.5	349.2
13	Feedwater						
14	flow	lbm/hr	474,318	472,121	472,133	474,353	473,231
15	pressure	psia	1410.2	1404.7	1404.6	1410.3	1407.5
16	temperature	°F	112.8	118.3	118.3	112.8	115.6
17	Economizer						
18	temperature	°F	554.5	553.9	553.9	554.5	554.2
19	Evaporator						
20	pressure	psia	1268.3	1264.1	1264.0	1268.4	1266.2
21	temperature	°F	574.2	573.8	573.8	574.3	574.0
22	Superheater						
23	pressure	psia	1204.7	1200.9	1200.8	1204.9	1202.8
24	temperature	°F	996.5	999.6	999.4	996.6	998.0

Steam properties used are the same ones in feedwater_heater_analytical.xls plus three additional functions have been provided as macros for the GT exhaust gas properties appropriate for this particular test. These could be calculated based on the information in the appendix of PTC 4.4 or from the NASA Glenn tables (NASA TP-2002-211556). The property section is:

	A	C	D	E	F	G	H
25	PROPERTIES	units					
26	GT Exhaust						
27	enthalpy	BTU/lbm	263.9	264.9	264.9	263.8	264.4
28	Stack						
29	enthalpy	BTU/lbm	72.7	73.1	73.1	72.7	72.9
30	Feedwater						
31	enthalpy	BTU/lbm	84.5	89.9	89.9	84.4	87.2
32	Economizer						
33	enthalpy	BTU/lbm	554.7	554.0	554.0	554.8	554.4
34	Evaporator						
35	enthalpy	BTU/lbm	1181.8	1182.0	1182.0	1181.8	1181.9
36	Superheater						
37	enthalpy	BTU/lbm	1497.2	1499.1	1499.0	1497.2	1498.1

The calculation section is:

38	CALCULATIONS	units					
39	Heat to Steam						
40	Economizer	10⁶BTU/hr	223.1	219.1	219.1	223.1	221.1
41	Evaporator	10⁶BTU/hr	297.4	296.5	296.5	297.4	297.0
42	Superheater	10⁶BTU/hr	149.6	149.7	149.7	149.6	149.7
43	Total	10⁶BTU/hr	670.1	665.3	665.3	670.2	667.7
44	Heat from Gas						
45	Total	10⁶BTU/hr	677.0	671.7	671.8	677.2	674.4
46	Loss	10⁶BTU/hr	6.9	6.4	6.5	7.0	6.7
47	Loss	%	1.03%	0.95%	0.96%	1.03%	0.99%
48	Gas Enthalpy						
49	Superheater	BTU/lbm	221.2	221.7	221.7	221.2	221.5
50	Evaporator	BTU/lbm	136.4	136.3	136.3	136.4	136.3
51	Gas Temp.						
52	Superheater	°F	910.0	912.0	912.1	910.0	911.0
53	Evaporator	°F	594.1	593.8	593.8	594.2	593.9
54	LMTDs						
55	Superheater	°F	167.3	168.6	168.8	167.1	167.9
56	Evaporator	°F	111.7	112.4	112.5	111.8	112.1
57	Economizer	°F	87.2	86.3	86.3	87.3	86.8
58	Heat Capacity						
59	Superheater						
60	Gas Side	10⁶BTU/hr/°F	0.981	0.971	0.971	0.982	0.976
61	Steam Side	10⁶BTU/hr/°F	0.354	0.352	0.352	0.354	0.353
62	R	-	0.361	0.362	0.362	0.361	0.362
63	P	-	0.311	0.312	0.312	0.311	0.312
64	F	-	0.991	0.991	0.991	0.991	0.991
65	Economizer						
66	Gas Side	10⁶BTU/hr/°F	0.918	0.907	0.907	0.918	0.913
67	Steam Side	10⁶BTU/hr/°F	0.908	0.899	0.898	0.908	0.903
68	R	-	0.990	0.990	0.990	0.990	0.990
69	P	-	0.505	0.508	0.508	0.505	0.506
70	F	-	0.843	0.838	0.838	0.843	0.841
71	UAs						
72	Superheater	10⁶BTU/hr/°F	0.903	0.896	0.895	0.904	0.900
73	Evaporator	10⁶BTU/hr/°F	2.662	2.637	2.636	2.661	2.649
74	Economizer	10⁶BTU/hr/°F	3.034	3.029	3.029	3.031	3.030
75	Total	10⁶BTU/hr/°F	6.598	6.562	6.561	6.596	6.579

All three heat exchangers are crossflow; however, the evaporator has a heat capacity ratio, R, of zero, simplifying the calculations. The superheater, evaporator, and economizer are analyzed separately. The uncertainty for each test period, if desired, would be calculated as before.

Chapter 9. Moisture Separator/Reheater

The steam generators in nuclear plant of the pressurized water reactor variety produce saturated steam at best. The point is, they don't produce superheated steam. Wet steam is very hard on a steam turbine. While steam turbines in a nuclear plant have special moisture extraction blades, this alone isn't enough. The steam must be somewhat dried and reheated. This is what a moisture separator/reheater (MSR) does.

A MSR looks like an oblong tank but contains several sections: a moisture separator plus one or two tube bundles. The low-pressure wet steam may enter at the top or bottom, but always flows through a stack of chevrons to remove water droplets, then upward across the tube bundle(s) to heat and dry what's left. The tube bundles are fed by high-pressure steam. Chevrons are depicted below:

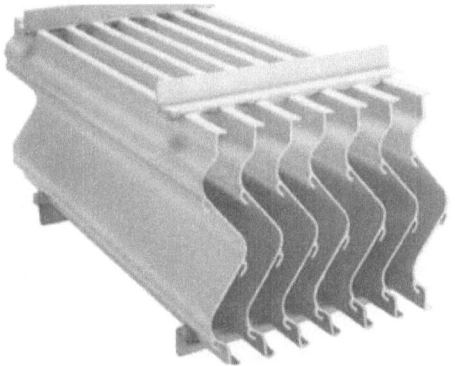

An MSR is not employed to improve thermal efficiency. On the contrary, it reduces thermal efficiency, sacrificing high-pressure steam to heat low-pressure steam, which generates entropy. The following is a typical schematic:

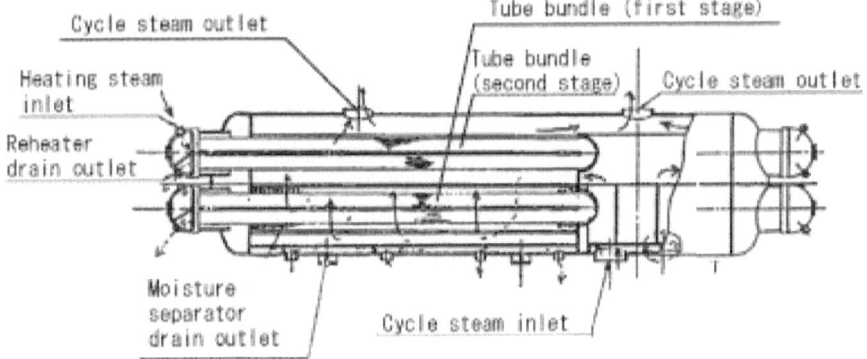

A simple analysis is presented here using the LMTD method. A more detailed, numerical example is presented in Chapter 14. The entering and exiting steam is most often not superheated; therefore, it is necessary to either specify enthalpy or quality. The pressure drop is also important and will be specified here as an input rather than calculating it, as in the more detailed example.

Performance Prediction

	A	B	C	D	E	F
1		Moisture Separator Reheater Example				
2	INPUTS	units	value	CALCS.	units	value
3	HP Tubes			LP Tubes		
4	Flow	lbm/hr	100,000	Tinlet	°F	443.6
5	Pinlet	psia	834.5	Toutlet	°F	438.7
6	Hinlet	BTU/lbm	1295.6	Houtlet	BTU/lbm	1054.9
7	Poutlet	psia	824.2	Shell Side		
8	Xoutlet	%	97%	Tinlet	°F	370.8
9	LP Tubes			Xinlet	%	88.7%
10	Flow	lbm/hr	75,000	Tchevr	°F	369.2
11	Pinlet	psia	396.1	Xchevr	%	98.3%
12	Hinlet	BTU/lbm	1150.1	Hchevr	BTU/lbm	1181.7
13	Poutlet	psia	376.4	Hmidl	BTU/lbm	1196.0
14	Xoutlet	%	81%	Tmidl	°F	368.7
15	Shell Side			Houtlet	BTU/lbm	1219.4
16	Flow	lbm/hr	500,000	Toutlet	°F	385.8
17	Pinlet	psia	175.0	superheat	°F	17.6
18	Hinlet	BTU/lbm	1100.0	LP Tubes		
19	chevrons	%	85%	Q	BTU/hr	7.14E+06
20	Pchevr	psia	171.7	LMTD	°F	72.2
21	Poutlet	psia	169.4	FUA	BTU/hr/°F	9.90E+04
22	CALCS.	units	value	HP Tubes		
23	HP Tubes			Q	BTU/hr	1.17E+07
24	Tinlet	°F	596.0	LMTD	°F	63.7
25	Toutlet	°F	521.7	FUA	BTU/hr/°F	1.84E+05
26	Houtlet	BTU/lbm	1178.2			

There is often condensation inside the tube bundles, which reduces the complexity and necessity of estimating F. In this simplified example, it will be left in the product, *FUA*. The most challenging problems facing the designer of MSRs are mechanical. The original design used in all of the Westinghouse® Pressurized Water Reactors (PWRs) had finned copper-nickel tubes. While these are great for heat transfer, they have a very large coefficient of thermal expansion and a low ultimate stress. This combination in a heat exchanger is a recipe for failure—in this case, a colossal one, costing millions of dollars.

In order to avoid these mechanical problems, condensation inside the tubes of this design must be minimized. Several approaches were taken, including flow-restricting orifice plates at the tube bundle outlet, but these were marginally successful at best. All of the MSRs of this original design have been replaced with improved designs having other materials (e.g., stainless or titanium) and sheets that allow the tubes to slide back-and-forth.

In addition to the removal of moisture droplets, superheating of the steam is of primary importance. The impacts of extraction steam flow rates on superheating and required FUA are shown in the next three figures.

Impact of Extraction Flow on Superheat										
	50,000	60,000	70,000	80,000	90,000	100,000	110,000	120,000	130,000	140,000	150,000
40,000	3.80	5.57	7.33	9.09	10.85	12.61	14.37	16.14	17.90	19.66	21.42
48,000	4.95	6.71	8.47	10.23	11.99	13.76	15.52	17.28	19.04	20.80	22.56
56,000	6.09	7.85	9.61	11.38	13.14	14.90	16.66	18.42	20.18	21.95	23.71
64,000	7.23	8.99	10.76	12.52	14.28	16.04	17.80	19.57	21.33	23.09	24.85
72,000	8.38	10.14	11.90	13.66	15.42	17.19	18.95	20.71	22.47	24.23	25.99
80,000	9.52	11.28	13.04	14.80	16.57	18.33	20.09	21.85	23.61	25.38	27.14
88,000	10.66	12.42	14.19	15.95	17.71	19.47	21.23	22.99	24.76	26.52	28.28
96,000	11.81	13.57	15.33	17.09	18.85	20.61	22.38	24.14	25.90	27.66	29.42
104,000	12.95	14.71	16.47	18.23	20.00	21.76	23.52	25.28	27.04	28.80	30.57
112,000	14.09	15.85	17.62	19.38	21.14	22.90	24.66	26.42	28.19	29.95	31.71
120,000	15.23	17.00	18.76	20.52	22.28	24.04	25.81	27.57	29.33	31.09	32.85

Impact of Extraction Flow on Required FUA										
	50,000	60,000	70,000	80,000	90,000	100,000	110,000	120,000	130,000	140,000	150,000
40,000	0.1357	0.1535	0.1718	0.1906	0.2099	0.2297	0.2502	0.2712	0.2929	0.3152	0.3384
48,000	0.1469	0.1649	0.1834	0.2023	0.2218	0.2419	0.2625	0.2838	0.3058	0.3284	0.3518
56,000	0.1582	0.1763	0.1950	0.2141	0.2338	0.2541	0.2750	0.2965	0.3187	0.3417	0.3654
64,000	0.1694	0.1877	0.2066	0.2259	0.2458	0.2663	0.2875	0.3092	0.3317	0.3550	0.3791
72,000	0.1807	0.1992	0.2182	0.2377	0.2579	0.2786	0.3000	0.3220	0.3448	0.3684	0.3928
80,000	0.1928	0.2116	0.2309	0.2507	0.2712	0.2923	0.3140	0.3365	0.3597	0.3837	0.4086
88,000	0.2058	0.2250	0.2446	0.2649	0.2857	0.3072	0.3294	0.3524	0.3761	0.4007	0.4262
96,000	0.2191	0.2386	0.2586	0.2793	0.3005	0.3225	0.3452	0.3687	0.3930	0.4182	0.4443
104,000	0.2326	0.2524	0.2729	0.2940	0.3157	0.3381	0.3613	0.3854	0.4102	0.4361	0.4629
112,000	0.2464	0.2666	0.2874	0.3090	0.3312	0.3541	0.3778	0.4024	0.4279	0.4544	0.4819
120,000	0.2604	0.2810	0.3023	0.3243	0.3470	0.3704	0.3947	0.4199	0.4460	0.4732	0.5015

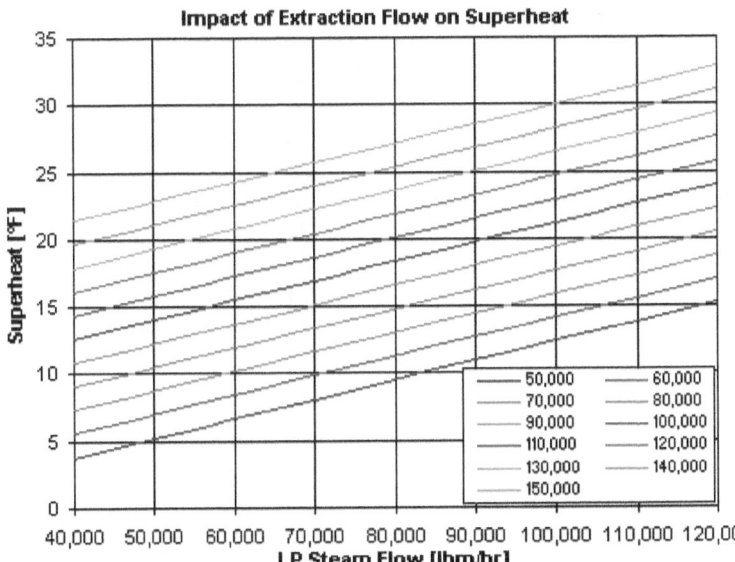

Impact of Extraction Flow on Superheat

47

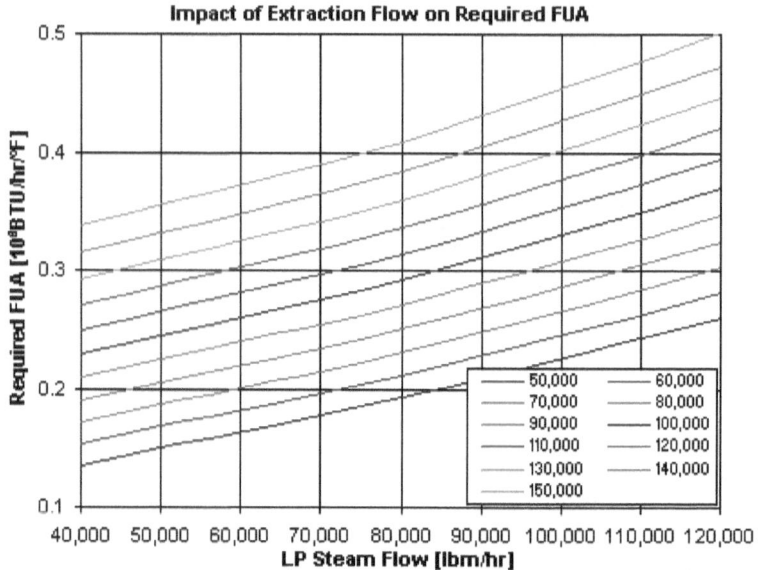

Performance Evaluation

It is virtually impossible to measure the thermal performance of a moisture separator/reheater. Not only are the flows two-phase (vapor plus liquid), but also the temperatures are not uniform across the flow stream. Besides being a useless exercise, traversing the flow with a probe is completely impractical. No one is going to let you stick a probe into anything inside an operating nuclear plant.

I can't say enough about the difficulties involved in measuring a two-phase flow. It is a known fact—thoroughly established by experiment—that the two phases travel at different speeds. Which one do you intend to measure? Presumably the vapor phase. Will you then calculate, estimate, or ignore the liquid phase? If so, why bother measuring anything? Resign yourself to calculating it from other things that you can measure; because that's as close as you will come.

Not only are the flows impractical to measure, so are the steam qualities. Even if you were to pass the steam through a calorimeter, you cannot assure that your sample is representative of the whole. Such a test would be pointless.

Chapter 10. The TEMA Designs

TEMA stands for Tubular Exchangers Manufacturers Association, a group of leading shell-and-tube designers and fabricators. TEMA has identified seven arrangements for the shell and tubes. TEMA has also identified a number of head/end types, which will not be discussed here, as these are mechanical considerations and we are only considering thermal ones. The seven arrangements are given letter designations (E, F, G, H, J, K, X) and are shown below:

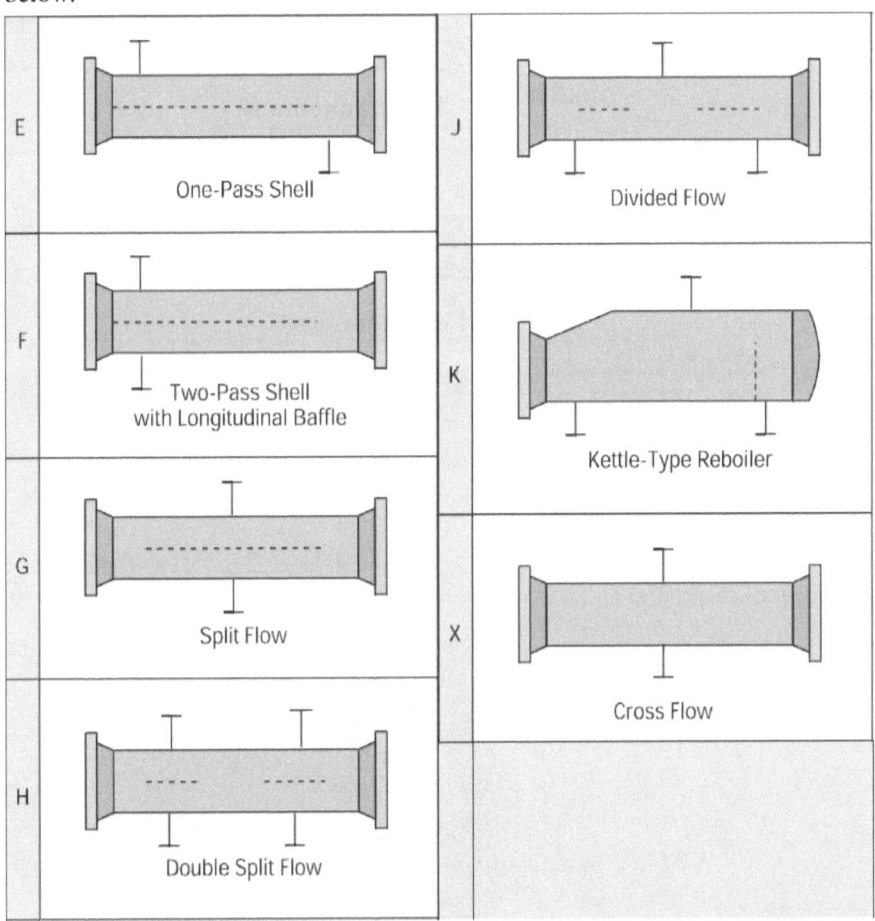

TEMA and its members have provided many resources to guide in the selection of the best design for most applications. These are readily available on the web. Tube-side and shell-side heat transfer coefficients for these designs are similar to those briefly discussed in Chapter 7. More details on the specifics (i.e., convective heat transfer inside tubes) may be found in any heat transfer textbook, such as the one referenced previously.

Performance Prediction

There are two ways of analyzing these designs: the F-LMTD/P-NTU method or numerically. The former will be presented here and the latter in subsequent chapters. Type-X is simply crossflow, which was covered in the previous chapter. Analytical solutions exist for many of these and separate solutions are required for variants, such as 1-2 and 1-4 arrangements, as illustrated in the following figure.

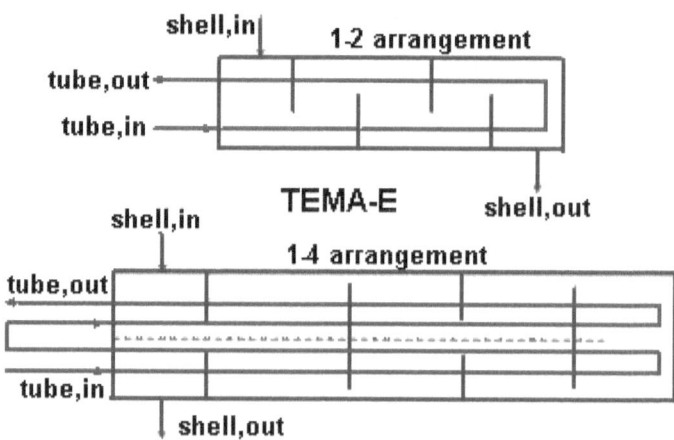

Only the 1-2 arrangement will be presented here. The NTU-ε method is readily applicable, and given in the following equations:

$$\varepsilon = \frac{2}{1+R+\sqrt{1+R^2}\left(\dfrac{1+e^{-\Gamma}}{1-e^{-\Gamma}}\right)} \quad (10.1)$$

$$\Gamma = NTU\sqrt{1+R^2} \quad (10.2)$$

$$NTU = \frac{1}{\sqrt{1+R^2}} \ln\left[\frac{2-\varepsilon\left(1+R-\sqrt{1+R^2}\right)}{2-\varepsilon\left(1+R+\sqrt{1+R^2}\right)}\right] \quad (10.3)$$

$$R = \frac{(\dot{m}C_P)_{MIN}}{(\dot{m}C_P)_{MAX}} \quad (10.4)$$

An illustration of this calculation may be found in the spreadsheet Eshell_analytical.xls, as shown below:

	A	B	C	D
1	TEMA E-Shell Example			
2	symbol	units	tube	shell
3	U	W/m²/°C	65	
4	Cp	kJ/kg/°C	2.5	1.0
5	m	kg/s	1000	8000
6	Tin	°C	250	50
7	Tout	°C	100	96.9
8	q	kW	375,000	
9	mCp	kW/°C	2,500	8,000
10	R	-	0.313	
11	ε	-	0.750	
12	NTU	-	1.965	
13	UA	-	4913	
14	A	m²	75.6	
15	user inputs in blue			
16	calculations in orange			

The NTU-ε method is used to calculate *NTU* and *UA*, the required conductance. It should be apparent by now that the product *UA* occurs quite frequently in thermal analyses of heat exchangers, much more so than either *U* or *A* alone. The overall conductance, *U*, is dominated by fluid and material properties as well as design and operation (i.e., Reynolds numbers, convective, evaporative, and condensive heat transfer coefficients). The heat exchange area, *A*, along with the overall design and selection of materials drives the cost.

Performance Evaluation

Evaluation of performance test data for this type of heat exchanger is the same as for the crossflow except for a few formulas in the calculation of **UA**. The uncertainty is calculated in exactly the same way.

TEMA E-Shell Heat Exchanger Test with Uncertainty									
measurement	units	average	points	std.dev.	bias	total unc.	sensitivity	units	contrib
Thot,in	°C	247.61	30	2.11	0.14	0.80	-34.2	kW/°C²	-27
Thot,out	°C	101.24	30	1.79	0.14	0.68	-59.4	kW/°C²	-41
Tcold,in	°C	51.53	30	1.65	0.14	0.63	-115.0	kW/°C²	-73
Tcold,out	°C	97.81	30	1.44	0.14	0.56	208.5	kW/°C²	116
m,hot	kg/s	1000	30	1.93%	2.0%	2.1%	-2.17	kWs/kg/°C	-46
m,cold	kg/s	8000	30	1.62%	2.0%	2.1%	0.889	kWs/kg/°C	149
calculation	units	value							213
mCp,hot	kW/°C	2500							4.3%
mCp,cold	kW/°C	8000							
R	-	0.313							
q,hot	kW	365,925							
q,cold	kW	370,240							
q,avg	kW	368,083							
ε,hot	-	0.746							
ε,cold	-	0.755							
NTU,hot	-	1.932							
NTU,cold	-	2.018							
UA,hot	kW/°C	4829							
UA,cold	kW/°C	5044							
UA,avg	kW/°C	4937							
UA,design	kW/°C	4913							
ΔUA	-	0.5%							

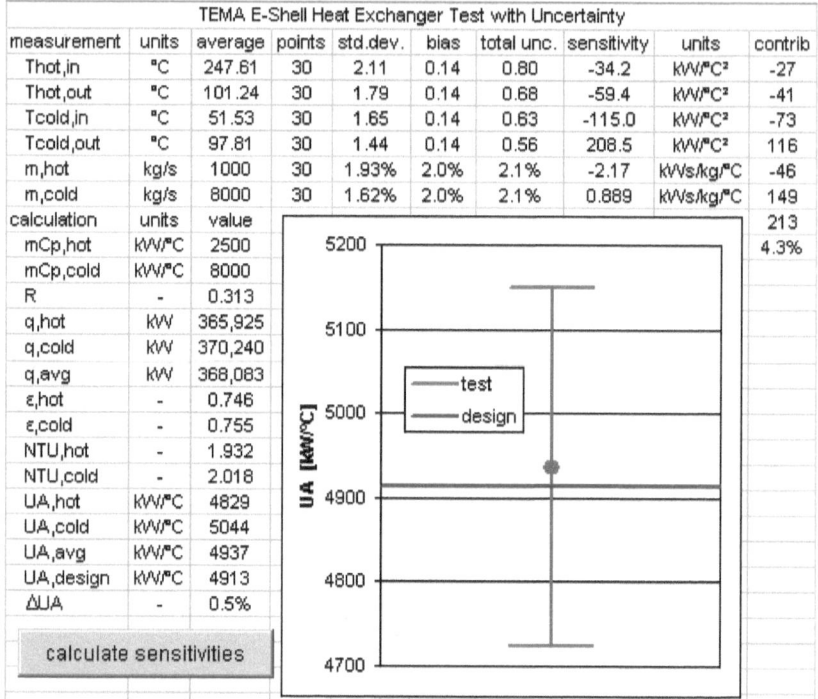

The analysis proceeds as for the crossflow example with instrument uncertainties, a button to calculate sensitivities, and an overall test result.

Chapter 11. Simple Numerical Methods

With the availability of personal computers there is little point developing complicated closed-form analytical solutions to such mundane problems. Numerical solutions can be easily set up in Excel® and used over and over again. These can be adapted to account for cross mixing, varying properties, and property-dependent heat transfer coefficients. The need for numerical solutions in such cases arises from the fact that the assumptions made in order to obtain the analytical solutions (e.g., the LMTD and P-NTU methods) are not valid.

We will begin with the simplest useful application: a crossflow arrangement without mixing. The heat exchanger is divided into cells. The number of cells required depends on how much the temperatures and properties vary. The strategy here is to simplify the problem so that the aforementioned assumptions are reasonably valid over a single cell, hoping that the combination of cells will accurately capture the whole.

> *Ensemble Hypothesis - the whole may be represented by an ordered assemblage of distinguishable, yet simpler parts, no one of which exhibits all the characteristics of the whole.*

In order to illustrate this approach, the first example problem in Chapter 3 will be divided into a 10x10 grid of cells. This and the following two examples may be found in crossflow_numerical.xls.

crossflow heat exchanger - fully explicit - 10x10 grid										
1724	hot side inlet										
	50	50	50	50	50	50	50	50	50	50
	34.9	40.1	43.5	45.8	47.2	48.2	48.8	49.2	49.5	49.7
	26.3	32.2	36.9	40.5	43.1	45.1	46.5	47.5	48.3	48.8
	21.4	26.5	31.1	35.1	38.5	41.2	43.3	45.0	46.3	47.3
	18.7	22.5	26.5	30.4	33.9	37.0	39.7	41.9	43.7	45.2
	17.1	19.8	23.0	26.4	29.8	33.0	35.9	38.5	40.7	42.6
	16.2	18.0	20.5	23.3	26.3	29.3	32.2	35.0	37.4	39.6
	15.7	16.9	18.7	20.9	23.4	26.1	28.9	31.6	34.2	36.5
	15.4	16.2	17.4	19.1	21.2	23.5	26.0	28.5	31.0	33.5
Texit	15.2	15.7	16.6	17.9	19.5	21.4	23.5	25.8	28.2	30.6
20.0	15.1	15.4	16.0	17.0	18.2	19.7	21.5	23.5	25.7	27.9
	hot side outlet										
15	27.1	35.0	40.2	43.6	45.8	47.2	48.2	48.8	49.2	49.5
15	21.9	28.2	33.5	37.7	41.0	43.5	45.3	46.7	47.6	48.3
15	18.9	23.5	28.1	32.4	36.1	39.2	41.7	43.7	45.3	46.5
15	17.2	20.4	24.1	27.9	31.6	34.9	37.8	40.3	42.4	44.1
15	16.3	18.4	21.2	24.4	27.7	30.9	33.9	36.7	39.1	41.2
15	15.7	17.1	19.2	21.7	24.5	27.4	30.3	33.1	35.7	38.1
15	15.4	16.3	17.8	19.7	22.0	24.5	27.2	29.9	32.5	34.9
15	15.2	15.8	16.8	18.2	20.0	22.1	24.5	26.9	29.4	31.9
15	15.1	15.5	16.2	17.2	18.6	20.3	22.2	24.4	26.7	29.0
15	15.1	15.3	15.8	16.5	17.5	18.8	20.5	22.3	24.3	26.5
								Texit	**39.0**	

(cold side inlet, cold side outlet)

In this case, the hot side temperatures form an 11x10 block (11 rows and 10 columns). There is an extra row at the top, as this serves as the hot side inlet boundary condition. The hot side exit temperature is the average of the bottom row in this block. The result (20.0) is shown in bold.

The cold side temperatures form a 10x11 block. There is an extra column on the left, as this serves as the cold side inlet boundary. The cold side exit temperature is the average of the right column in this block. The result (39.0) is shown in bold.

This first example is a fully explicit temperature difference, that is, using only the hot and cold temperatures entering each cell, in this case, above and to the left, respectively.

$$\Delta T_{i,j} = (T_H)_{i,j} - (T_C)_{i,j} \qquad (11.1)$$

This is implemented in the spreadsheet by typing =MAX(0,G3-F15) into cell R4 and dragging it to fill the block. MAX() is used to prevent overshoot. The temperature differences and the heat transfer each form a 10x10 block:

ΔT in each cell										
use max(0,dT) to prevent overshoot										
35.0	22.9	15.0	9.8	6.4	4.2	2.8	1.8	1.2	0.8
19.9	18.2	15.4	12.3	9.5	7.2	5.3	3.9	2.8	2.0
11.3	13.3	13.4	12.3	10.7	9.0	7.3	5.8	4.6	3.5
6.4	9.3	10.7	11.0	10.6	9.6	8.5	7.2	6.0	4.9
3.7	6.2	8.1	9.2	9.6	9.4	8.8	8.0	7.0	6.1
2.1	4.1	5.9	7.3	8.1	8.5	8.5	8.1	7.5	6.8
1.2	2.6	4.2	5.5	6.6	7.4	7.8	7.8	7.6	7.1
0.7	1.7	2.9	4.1	5.2	6.1	6.8	7.1	7.2	7.1
0.4	1.1	2.0	3.0	4.0	4.9	5.7	6.3	6.6	6.8
0.2	0.7	1.3	2.1	3.0	3.9	4.7	5.4	5.9	6.2
ΔQ in each cell										
603	395	259	170	111	73	48	31	20	13
343	315	265	212	164	124	92	67	49	35
195	230	231	213	185	155	126	100	78	60
111	160	185	190	182	166	146	124	104	85
63	108	140	158	165	162	152	137	121	104
36	71	102	125	140	147	146	140	130	117
20	45	72	96	114	127	134	134	131	123
12	29	50	71	90	106	117	123	125	123
7	18	34	51	69	85	99	108	114	117
4	11	22	36	52	67	81	92	101	107
								q	**12,000**
								error	0.0%

If Nx is the number of cells perpendicular to the hot stream (in the direction of the cold stream) and Ny is the number of cells perpendicular to the cold stream (in the direction of the hot stream) then the heat transfer in each cell is simply:

$$Q_{i,j} = \frac{UA}{NxNy} \Delta T_{i,j} \qquad (11.2)$$

This is implemented by typing =$E$2*R4/10/10 into cell R15 and dragging it to fill the block. The temperature change of the cold stream in a cell is:

$$(T_C)_{i,j+1} = (T_C)_{i,j} + \frac{Q_{i,j}}{\left(\dfrac{\dot{m}_C}{Ny}\right) C_{PC}} \qquad (11.3)$$

This is implemented by typing =F15+R15/$D$4/($D$5/10) into cell G15 and dragging it to fill the block. The temperature change of the hot stream in a cell is:

$$(T_H)_{i+1,j} = (T_H)_{i,j} - \frac{Q_{i,j}}{\left(\dfrac{\dot{m}_H}{Nx}\right) C_{PH}} \qquad (11.4)$$

This is implemented by typing =G3-R15/$C$4/($C$5/10) into cell G4 and dragging it to fill the block. The total heat transfer is the sum of all $Q_{i,j}$ and is shown in bold (12,000). The required conductance, *UA*, is placed in cell E2 and shown in orange (1724) and is assumed constant throughout. This requires iterative solution and a button is provided to accomplish this.

<u>How many grid cells is enough?</u>

The answer is not as simple as it might seem. First we increase the number of grid cells to see if this changes the result. It's not convenient to do this in Excel®, so a simple program written in C is used that dynamically allocates memory and runs much faster than a spreadsheet. The code is described in Appendix A and is included in the on-line archive along with the spreadsheet.

The result converges to 2800 kW/°C at about 400x400=160,000 cells (not something you'd want to try in Excel®). The analytical answer is 2362, which occurs at 33x33=1089 cells. Sometimes having too many cells leads to round-off errors, but this is not what's happening here. The problem is more subtle than this.

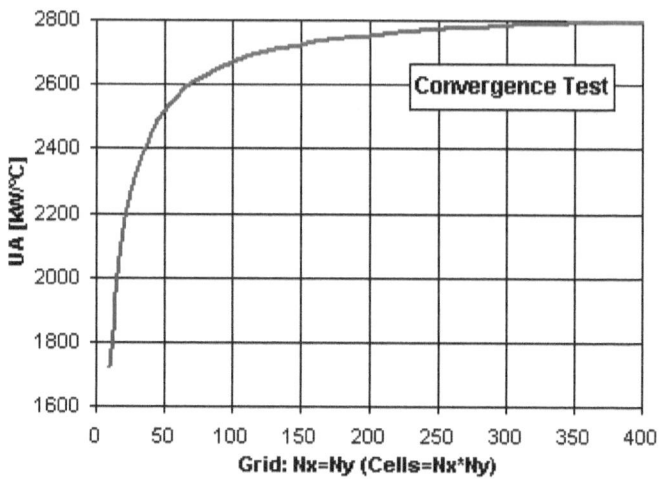

Explicit vs. Implicit Differences

Explicit temperature differences were used in this example up until this point. This over-estimates the thermal driving potential with a coarse grid. There are many possible corrections to the fully explicit difference that might be considered: average (arithmetic mean), log-mean, geometric mean, and harmonic mean.

Using only explicit temperature differences, it is possible to sweep through the grid in a single pass. This is not possible with implicit temperature differences, because the calculations for each cell depend on the results. This produces circular references. If you type =AVERAGE(G3-F15,G4-G15) into cell R4 and drag to fill the block, Excel® will display an error and blue arrows showing the circular references. You can get past this by enabling iterative calculations (see Tools/Options or File/Options/Formulas). This modification will slow the calculations considerably, but will significantly improve the accuracy of the results. The impact of this modification is shown in the next figure.

The 10x10 grid converges after 7 or 8 iterations. The initial hump at 2 or 3 iterations is overshoot and to be avoided. For grid sizes above about 15x15, this simple modification is not adequate, as it yields the same erroneous overshot solution. For our purposes here, a 10x10 grid with 8 implicit iterations of two-thirds/one-third weighting is adequate. Excel® handles the iterations automatically, as long as it doesn't get too far off, producing #VALUEs.

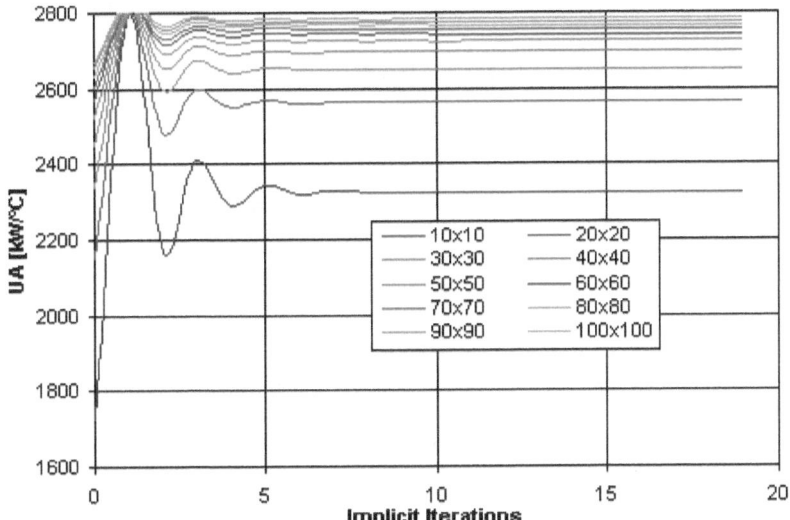

While this modification works well enough, it requires iterations and results in circular references in Excel®. The most efficient modification is to use the geometric mean, which has a closed-form solution. In terms of the cell indices, this becomes:

$$\Delta T_{i,j} = \sqrt{(\Delta T_A)(\Delta T_B)}$$
$$\Delta T_A = (T_H)_{i,j} - (T_C)_{i,j} \quad (11.5)$$
$$\Delta T_B = (T_H)_{i+1,j} - (T_C)_{i,j+1}$$

Substituting Equation 11.5 into 11.2 and solving 11.3 and 11.4 for the hot side exit temperature yields:

$$(T_H)_{i+1,j} = (T_H)_{i,j} + \frac{\Delta T_A \beta [(Ny + \alpha Nx)\beta - \sqrt{\gamma}]}{\delta Ny} \quad (11.6)$$

$$\alpha = \frac{\dot{m}_C C_{PC}}{\dot{m}_H C_{PH}} \quad (11.7)$$

$$\beta = \frac{UA}{\dot{m}_H C_{PH}} \quad (11.8)$$

$$\delta = \frac{2\alpha}{NxNy} \quad (11.9)$$

$$\gamma = (\alpha Nx^2 + Ny^2 + \delta)\beta^2 + \delta^2 \quad (11.10)$$

Not only does the geometric mean lead to a closed-form (non-iterative) solution, it supports a much coarser grid size than the arithmetic mean. Convergence with grid size is quite remarkable compared to the previous figure, which is why this adaptation is used whenever possible.

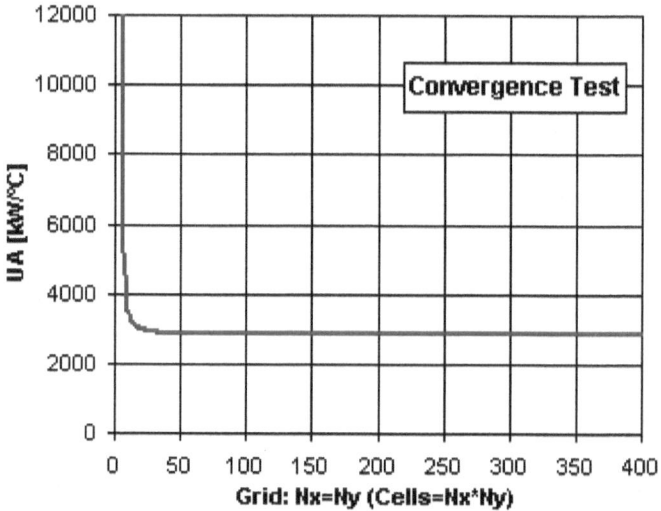

Chapter 12: Variable Properties

Fluid properties often vary only slightly throughout a heat exchanger and can be adequately approximated by average values; however, this is not always the case. The following extreme example of a simple double-pipe counterflow arrangement will be used to illustrate the potential difficulty of analyzing heat exchanger performance with significantly varying properties.

	A	B	C	D	E	F	G	H	I	J
1	Extreme Variable Property Example										
2	Inputs		Q	Th	Tc	dT	1/dT	CpH	CpC	dQ
3	mH	1000	0%	300.0	100.0	200.0	0.0050	6.031	0.904	14999
4	mC	1000	1%	302.5	116.6	185.9	0.0054	5.347	0.923	13942
5	UA	7500	2%	305.1	131.7	173.4	0.0058	4.820	0.941	13005
6	Q	6.26E+05	3%	307.8	145.5	162.3	0.0062	4.405	0.959	12171
7	LMTD method		4%	310.6	158.2	152.4	0.0066	4.071	0.976	11427
8	LMTD	200	5%	313.4	169.9	143.5	0.0070	3.798	0.993	10759
9	UA	3126	6%	316.2	180.7	135.5	0.0074	3.572	1.009	10159
10	error	-58.3%	7%	319.0	190.8	128.2	0.0078	3.381	1.025	9617
11	NTU-ε method		8%	321.9	200.2	121.7	0.0082	3.218	1.041	9127
12	CpH,avg	1.836	9%	324.7	209.0	115.8	0.0086	3.077	1.056	8682
13	CpC,avg	1.893	10%	327.5	217.2	110.4	0.0091	2.954	1.071	8277
14	mCh	1836	11%	330.3	224.9	105.4	0.0095	2.846	1.085	7907
15	mCc	1893	12%	333.1	232.2	100.9	0.0099	2.750	1.100	7569
16	R	0.970	13%	335.9	239.1	96.8	0.0103	2.665	1.114	7259
17	ε	0.667	14%	338.6	245.6	93.0	0.0108	2.587	1.128	6975
18	NTU	1.942	15%	341.3	251.8	89.5	0.0112	2.517	1.141	6713
19	UA	3566	16%	344.0	257.7	86.3	0.0116	2.454	1.155	6472
20	error	-52.4%	17%	346.6	263.3	83.3	0.0120	2.395	1.168	6249
21	Numerical method		18%	349.2	268.6	80.6	0.0124	2.341	1.181	6044
22	ΔTmean	76.7	19%	351.8	273.7	78.1	0.0128	2.292	1.194	5854
23	UA	8162	99%	684.5	497.1	187.4	0.0053	0.906	5.559	14054
24	error	8.8%	100%	700.0	499.6	200.4		0.889	6.257	

The units are immaterial and have been left blank intentionally. In this example, the specific heat of both fluids varies from inlet to exit by a factor of 7. This carefully selected variation in specific results in curvature of the T vs. Q curves, invalidating the assumptions made in developing both the LMTD and NTU-ε methods.

The T vs. Q curves are far apart at the inlet and exit (i.e., large ΔT) and close together in the middle (i.e., large ΔT). Simply averaging the specific heats over the length of the heat exchanger will not account for this variable temperature difference. In fact, there is no way to evaluate this process other than by numerically integrating or employing a finite difference or finite element technique, which would accomplish the same thing.

The LMTD and NTU-ε methods in this case have -58.3% and -50.7% errors, respectively. Even the numerical solution is off by 8.8% with 100 cells.

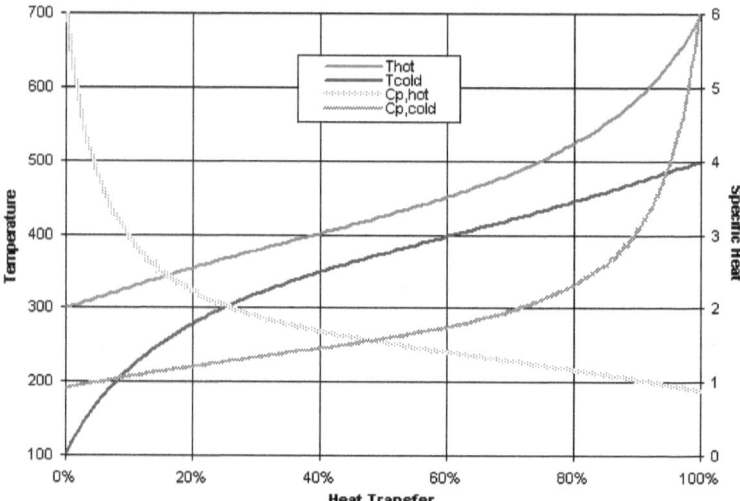

The geometric mean temperature difference is the only one that comes close to the correct solution with only a 3.2% error in overall conductance, **UA**. In this case the geometric mean is calculated by:

$$\Delta Tmean = \left[\prod_{i=1}^{n} \Delta T_i\right]^{\frac{1}{n}} \tag{12.1}$$

	A	B	C	D	E	F	G	H	I	J
1	Moderately Variable Property Example									
2	Inputs		Q	Th	Tc	dT	1/dT	CpH	CpC	dQ
3	mH	1000	0%	300.0	100.0	200.0	0.0050	1.96	1.24	22333
4	mC	1000	1%	311.4	118.0	193.4	0.0052	2.01	1.29	21594
5	UA	11167	2%	322.1	134.7	187.4	0.0053	2.06	1.34	20927
6	Q	1.22E+06	3%	332.3	150.3	182.0	0.0055	2.11	1.39	20320
7	LMTD method		4%	341.9	164.9	177.0	0.0056	2.15	1.44	19766
8	LMTD	129	5%	351.1	178.7	172.5	0.0058	2.20	1.49	19258
9	UA	9456	6%	359.9	191.6	168.3	0.0059	2.24	1.53	18789
10	error	-15.3%	7%	368.3	203.9	164.4	0.0061	2.28	1.57	18356
11	NTU-ε method		8%	376.4	215.6	160.8	0.0062	2.32	1.62	17953
12	CpH,avg	3.386	9%	384.1	226.7	157.4	0.0064	2.36	1.66	17578
13	CpC,avg	2.764	10%	391.5	237.3	154.3	0.0065	2.40	1.70	17227
14	mCh	3386	11%	398.7	247.4	151.3	0.0066	2.43	1.74	16898
15	mCc	2764	12%	405.7	257.1	148.6	0.0067	2.47	1.78	16589
16	R	0.816	13%	412.4	266.4	145.9	0.0069	2.51	1.82	16297
17	ε	0.817	14%	418.9	275.4	143.5	0.0070	2.54	1.85	16022
18	NTU	3.257	15%	425.2	284.1	141.2	0.0071	2.57	1.89	15762
19	UA	9002	16%	431.3	292.4	138.9	0.0072	2.61	1.93	15515
20	error	-19.4%	17%	437.3	300.4	136.8	0.0073	2.64	1.96	15280
21	Numerical method		18%	443.1	308.2	134.8	0.0074	2.67	2.00	15057
22	ΔTmean	105.3	19%	448.7	315.8	132.9	0.0075	2.70	2.03	14844
23	UA	11545	99%	698.0	621.2	76.9	0.0130	4.35	3.79	8582
24	error	3.4%	100%	700.0	623.4	76.6		4.36	3.80	

60

This second illustration has a more moderate variation of properties. The LMTD method is now only -15.3% off and the NTU-ε method is -19.4% off. The numerical solution is off by 3.4% with 100 cells. The T vs. Q curves are much closer to being typical, though still significantly curved.

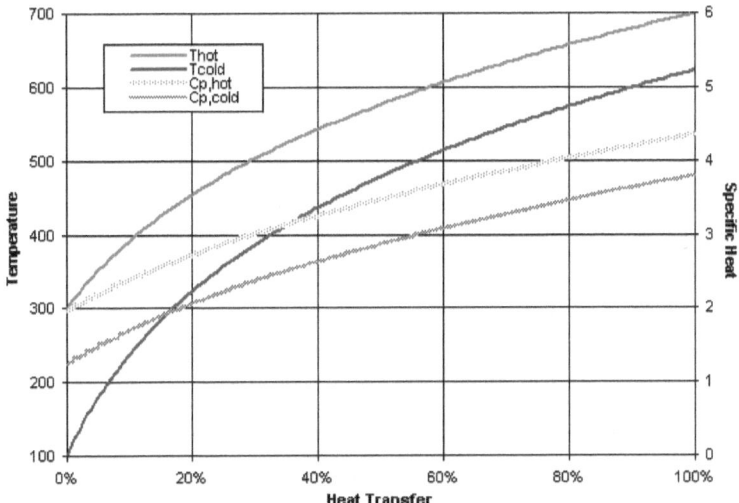

Any of these methods will work in the case of constant properties:

	A	B	C	D	E	F	G	H	I	J
1				Contstant Property Example						
2	Inputs		Q	Th	Tc	dT	1/dT	CpH	CpC	dQ
3	mH	1000	0%	300.0	100.0	200.0	0.0050	2.50	3.50	7928
4	mC	1000	1%	303.2	102.3	200.9	0.0050	2.50	3.50	7964
5	UA	3964	2%	306.4	104.5	201.8	0.0050	2.50	3.50	8000
6	Q	1.00E+06	3%	309.6	106.8	202.7	0.0049	2.50	3.50	8036
7	LMTD method		4%	312.8	109.1	203.6	0.0049	2.50	3.50	8072
8	LMTD	253	5%	316.0	111.4	204.6	0.0049	2.50	3.50	8109
9	UA	3955	6%	319.2	113.7	205.5	0.0049	2.50	3.50	8146
10	error	-0.2%	7%	322.5	116.1	206.4	0.0048	2.50	3.50	8182
11	NTU-ε method		8%	325.8	118.4	207.4	0.0048	2.50	3.50	8220
12	CpH,avg	2.500	9%	329.1	120.8	208.3	0.0048	2.50	3.50	8257
13	CpC,avg	3.500	10%	332.4	123.1	209.2	0.0048	2.50	3.50	8294
14	mCh	2500	11%	335.7	125.5	210.2	0.0048	2.50	3.50	8332
15	mCc	3500	12%	339.0	127.9	211.1	0.0047	2.50	3.50	8369
16	R	0.714	13%	342.4	130.3	212.1	0.0047	2.50	3.50	8407
17	ε	0.667	14%	345.7	132.7	213.1	0.0047	2.50	3.50	8445
18	NTU	1.582	15%	349.1	135.1	214.0	0.0047	2.50	3.50	8484
19	UA	3955	16%	352.5	137.5	215.0	0.0047	2.50	3.50	8522
20	error	-0.2%	17%	355.9	139.9	216.0	0.0046	2.50	3.50	8561
21	Numerical method		18%	359.3	142.4	217.0	0.0046	2.50	3.50	8600
22	ΔTmean	250.1	19%	362.8	144.8	217.9	0.0046	2.50	3.50	8639
23	UA	3998	99%	695.0	382.2	312.9	0.0032	2.50	3.50	12402
24	error	0.9%	100%	700.0	385.7	314.3		2.50	3.50	

The -0.2%, -0.2%, and 0.9% errors for the LMTD, NTU-ε, and numerical methods, respectively, are due to cumulative round-off and not reflective of any shortcomings in these analyses.

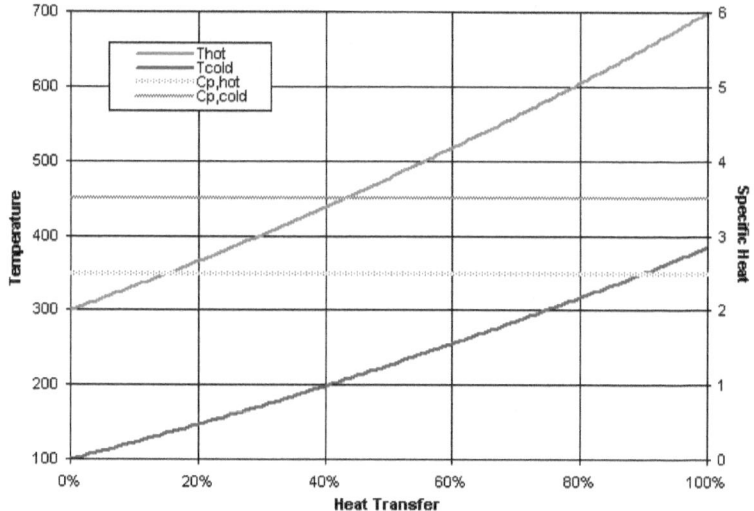

Chapter 13: Variable Conductance

Heat transfer coefficients may also vary within a heat exchanger leading to a variable overall conductance, UA_{local}. The impact of this is quite different than for variable properties because of how the governing differential equation was separated and integrated. In this first example, U varies is with ΔT^n, $n<0$.

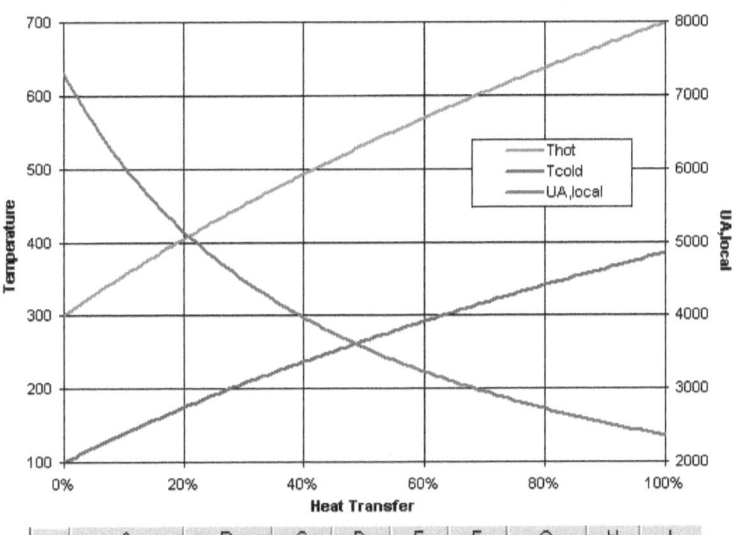

	A	B	C	D	E	F	G	H	I
1	Variable Conductance Example 1: UA,local=α*(ΔT/100)^n, n<0								
2	Inputs		Q	Th	Tc	ΔT	1/ΔT	UA	dQ
3	CpH	2.5	0%	300.0	100.0	200.0	0.0050	7310	14619
4	CpC	3.5	1%	305.8	104.2	201.7	0.0050	7159	14438
5	mH	1000	2%	311.6	108.3	203.3	0.0049	7015	14262
6	mC	1000	3%	317.3	112.4	205.0	0.0049	6876	14093
7	n	-2.5	4%	323.0	116.4	206.6	0.0048	6743	13928
8	α	41349	5%	328.5	120.4	208.2	0.0048	6615	13769
9	Q	1.00E+06	6%	334.0	124.3	209.7	0.0048	6491	13614
10	LMTD method		7%	339.5	128.2	211.3	0.0047	6372	13464
11	LMTD	253	8%	344.9	132.1	212.8	0.0047	6258	13318
12	UA	3955	9%	350.2	135.9	214.3	0.0047	6147	13177
13	error	3.2%	10%	355.5	139.6	215.8	0.0046	6041	13039
14	NTU-ε method		11%	360.7	143.3	217.3	0.0046	5938	12905
15	mCh	2500	12%	365.8	147.0	218.8	0.0046	5838	12775
16	mCc	3500	13%	371.0	150.7	220.3	0.0045	5742	12648
17	R	0.714	14%	376.0	154.3	221.7	0.0045	5649	12524
18	ε	0.667	15%	381.0	157.9	223.2	0.0045	5559	12404
19	NTU	1.582	16%	386.0	161.4	224.6	0.0045	5471	12287
20	UA	3955	17%	390.9	164.9	226.0	0.0044	5387	12173
21	error	3.2%	18%	395.8	168.4	227.4	0.0044	5305	12061
22	Numerical method		19%	400.6	171.9	228.7	0.0044	5225	11952
23	ΔTmean	260.8	20%	405.4	175.3	230.1	0.0043	5148	11846
24	UA,mean	3834	21%	410.1	178.7	231.5	0.0043	5073	11742

The temperature difference increases from left to right (i.e., the distance between the red and blue curves) and the local conductance decreases (i.e., the falling brown line). The LMTD and NTU-ε methods yield the same solution, off by only 3.2%. In this second example $n>0$.

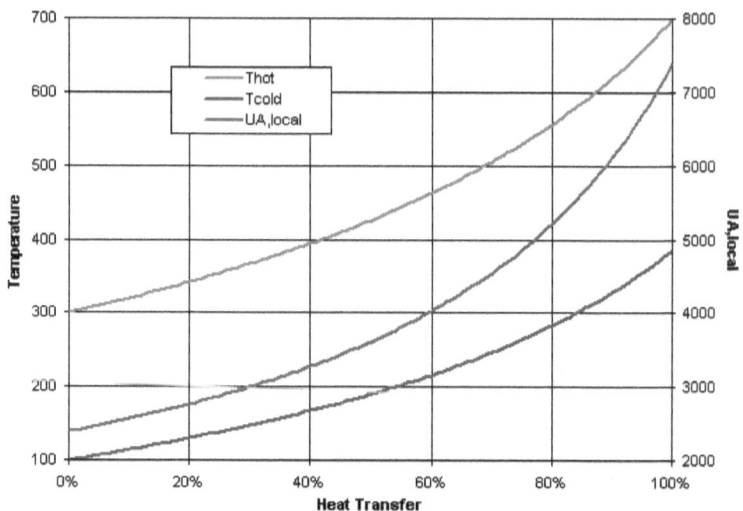

	A	B	C	D	E	F	G	H	I
1	Variable Conductance Example 2: UA,local=α*(ΔT/100)^n, n>0								
2	Inputs		Q	Th	Tc	ΔT	1/ΔT	UA	dQ
3	CpH	2.5	0%	300.0	100.0	200.0	0.0050	2388	4776
4	CpC	3.5	1%	301.9	101.4	200.5	0.0050	2404	4822
5	mH	1000	2%	303.8	102.7	201.1	0.0050	2421	4869
6	mC	1000	3%	305.8	104.1	201.7	0.0050	2438	4916
7	n	2.5	4%	307.8	105.5	202.2	0.0049	2455	4964
8	α	422	5%	309.7	107.0	202.8	0.0049	2472	5013
9	Q	1.00E+06	6%	311.7	108.4	203.4	0.0049	2490	5063
10	LMTD method		7%	313.8	109.8	203.9	0.0049	2507	5113
11	LMTD	253	8%	315.8	111.3	204.5	0.0049	2525	5165
12	UA	3955	9%	317.9	112.8	205.1	0.0049	2544	5217
13	error	-5.1%	10%	320.0	114.3	205.7	0.0049	2562	5270
14	NTU-ε method		11%	322.1	115.8	206.3	0.0048	2581	5325
15	mCh	2500	12%	324.2	117.3	206.9	0.0048	2600	5380
16	mCc	3500	13%	326.4	118.8	207.5	0.0048	2619	5436
17	R	0.714	14%	328.5	120.4	208.2	0.0048	2639	5493
18	ε	0.667	15%	330.7	121.9	208.8	0.0048	2659	5551
19	NTU	1.582	16%	332.9	123.5	209.4	0.0048	2679	5611
20	UA	3955	17%	335.2	125.1	210.1	0.0048	2700	5671
21	error	-5.1%	18%	337.5	126.8	210.7	0.0047	2721	5732
22	Numerical method		19%	339.8	128.4	211.4	0.0047	2742	5795
23	ΔTmean	240.0	20%	342.1	130.1	212.0	0.0047	2763	5859
24	UA,mean	4166	21%	344.4	131.7	212.7	0.0047	2785	5924

The LMTD and NTU-ε methods also yield the same solution in this case, off by only -5.1%. Variable *UA* has so much less dramatic impact on the analysis because the heat transfer is directly proportional it. While the heat transfer, Q, is also proportional to ΔT, ΔT decreases with increasing Q, so that ΔT is a function of itself.

Chapter 14. Two-Phase Flow Inside Tubes

The moisture separator/reheater presented in Chapter 9 is a real beast to analyze for the reasons already mentioned. In this chapter we will consider another reason: two-phase flow inside the tubes. There are several excellent texts on two-phase flow[10,11,12]. Collier in Section 2.5.2 describes Chisholm's Method[13] for pressure drop in horizontal pipes, which will be used here.

We begin with Lockart-Martinelli factor characterizing the two-phase flow.[14] This involves several parameters, including:

$$X^2 = \frac{\left(\frac{dp}{dx}\right)_f}{\left(\frac{dp}{dx}\right)_g} = \frac{\phi_g^2}{\phi_f^2} \qquad (14.1)$$

Here $dp/dx)_f$ and $dp/dx)_g$ are the pressure gradients due to friction for the liquid and vapor phases, respectively. The ratio of these two yields the square of empirical factor X. The two terms, φ_f and φ_g, are called the two-phase friction multipliers. It is furthermore assumed that these are related in the following form:

$$\phi_f^2 = 1 + \frac{C}{X} + \frac{1}{X^2} \qquad (14.2)$$

$$\phi_g^2 = 1 + CX + X^2 \qquad (14.3)$$

where C is a constant depending only on the combination of the two single-phase flow regimes given in the following table:

liquid	vapor	C
turbulent	turbulent	20
laminar	turbulent	12
turbulent	laminar	10
laminar	laminar	5

---

[10] Collier, J. G., *Convectivve Boiling and Condensation*, McGraw-Hill, 1972.
[11] Hsu, Y.-Y., and R. W. Graham, *Transport Processes in Boiling and Two-Phase Systems*, Hemisphere, 1976.
[12] Tong, L. S., *Boiling Heat Transfer and Two-Phase Flow*, Krieger, 1975.
[13] Chisholm, D., "The Influence of Mass Velocity on Friction Pressure Gradients During Steam-Water Flow," Paper 35, Thermodynamics and Fluid Mechanics Convention I Mech. Engrs. Bristol, 1968.
[14] Lockhart, R. W., and R. C. Martinelli, "Proposed Correlation of Data for Isothermal Two-Phase Two-Component Flow in Pipes," Chem. Eng. Prog., Vol. 45, No. 39, 1949.

The two-phase friction multipliers are shown in the following figure. The liquid multipliers start high and decrease with increasing X and the vapor ones do the opposite.

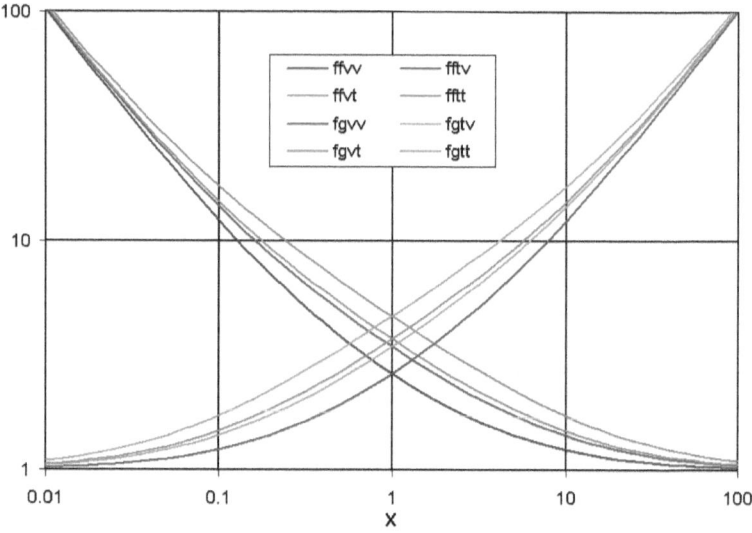

Colebrook-White[15] formula is used to calculate the single-phase friction factor for flow in pipes:

$$\frac{1}{\sqrt{f}} = -2\log_{10}\left(\frac{\varepsilon}{3.7D_h} + \frac{2.51}{\text{Re}\sqrt{f}}\right) \tag{14.4}$$

where D_h is the hydraulic diameter and Re is the Reynolds number. Chisholm modified this method, defining the coefficient, C, in the following manner:

$$C = \left[\lambda + (C_2 - \lambda)\left(\frac{v_{fg}}{V_g}\right)\right]\left[\sqrt{\frac{v_g}{v_f}} + \sqrt{\frac{v_f}{v_g}}\right] \tag{14.5}$$

$$\lambda = \frac{\left[2^{(2-n)} - 2\right]}{2} \tag{14.6}$$

$$C_2 = \frac{G*}{G} \tag{14.7}$$

---

[15] Colebrook, C. F. and White, C. M., "Experiments with Fluid Friction in Roughened Pipes". Proceedings of the Royal Society of London, Series A, Mathematical and Physical Sciences, Vol. 161, No. 906, pp. 367–381, 1937.

where v_f and v_g are the saturated liquid and vapor specific volumes, respectively, and v_{fg} is the difference between the two. For rough pipes $n=0$ and for smooth pipes $n=0.25$. The ratio, C_2, is reference mass flux, G^*, over the actual mass flux, G. For smooth tubes $G^*=2000$ kg/s/m² (1.47x10⁶ lb/hr/ft²) and for rough tubes $G^*=1500$ kg/s/m² (1.1x10⁶ lb/hr/ft²). Zivi's kinetic energy model[16] is used for the void ratio, α, is the ratio of the vapor to total areas in the pipe:

$$\alpha = \frac{A_g}{A} = \frac{1}{1 + \left(\frac{1-x}{x}\right)\left(\frac{\rho_g}{\rho_f}\right)^{\frac{2}{3}}} \tag{14.8}$$

The slip ratio is the ratio of the vapor to the liquid velocities:

$$S = \frac{u_g}{u_f} = \left(\frac{\rho_g}{\rho_f}\right)^{\frac{1}{3}} \tag{14.9}$$

The two-phase flow regime is given by the model of Breber, Palen, and Taborek.[17] The conceptual regimes are illustrated in this first figure:

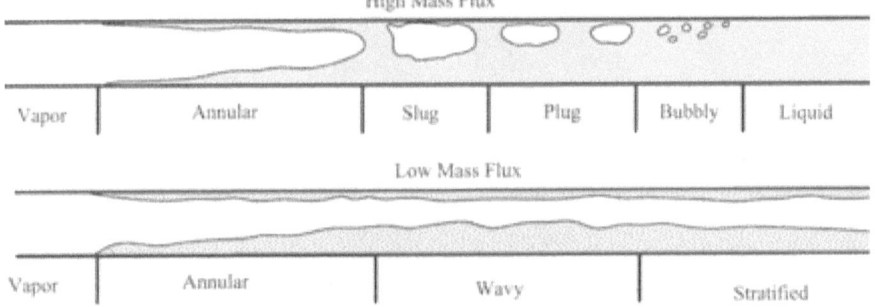

An excellent reference is the
Engineering Data Book III
by John R. Thome, published by
Wolverine Tube, Inc. and available on-line at:
https://pdfs.semanticscholar.org/9edc/9fe6557a5ecd15d7d773836a95fefa940c33.pdf

---

[16] Zivi, S.M., "Estimation of Steady State Steam Void Fraction by Means of the Principle of Minimum Entropy Production," ASME Journal of Heat Transfer, Vol. 86, pp. 247-252, 1964.

[17] Breber, G., J. W. Palen, and J. Taborek, "Prediction of Horizontal Tubeside Condensation of Pure Components Using Flow Regime Criteria," ASME Journal of Heat Transfer, Vol. 102, pp. 471-476, 1980.

The empirical relationship is illustrated in this second figure:

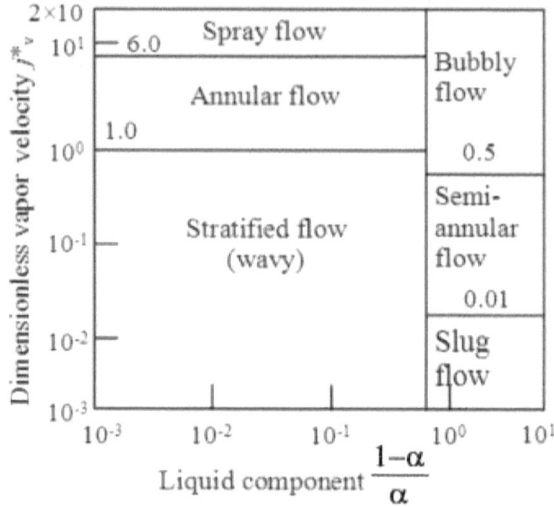

where the dimensionless vapor velocity, j^*_v, is given by:

$$j^*_v = \frac{x\dot{m}}{\sqrt{gD\rho_g(\rho_f - \rho_g)}} \qquad (14.10)$$

Once the flow regime is known, the heat transfer coefficient can be calculated using the Taitel-Dukler method[18]. The frictional pressure drop on the shell side is calculated using the method of Robinson and Briggs.[19] The frictional pressure drop on the shell side is calculated using the method of Briggs and Young.[20] It takes all of these steps just to calculate the pressure drop and heat transfer inside a single computational element along the pipe. This entire formulation has been written into a computer code that is listed in Appendix B.

Only select parts of the program output will be shown here. The complete output is included in the on-line archive. There are two banks of U-tubes: the high-pressure (higher temperature) at the top plus the low-pressure (lower

---

[18] Taitel and A.E. Dukler, "A Model for Predicting Flow Regime Transitions in Horizontal and Near Horizontal Gas-Liquid Flow", AIChE Journal, Vol. 22, No. 1, pp. 47–55, 1976.

[19] Robinson, K., and D. E. Briggs, "Pressure Drop of Air Flowing Across Triangular Pitch Banks of Finned Tubes," Eighth National Heat Transfer Conference, pp. 177-184, 1965.

[20] Briggs, D.E., and E.H. Young, "Convection Heat Transfer and Pressure Drop of Air Flowing Across Triangular Pitch Banks of Finned Tubes," Chem. Eng. Prog. Symp. Ser., Vol. 59, No. 41, pp. 1–10, 1963.

temperature) at the bottom. As these are U-tubes, the flow starts at the left, extends to the right, turns around, and flows back to the left. This first figure shows the flow patterns:

```
                    TWO-PHASE FLOW PATTERS INSIDE TUBES
 1H>  ENTR SUPR SUPR SUPR SUPR SUPR SUPR SUPR SUPR SUPR SUPR SUPR SUPR SUPR SUPR
 2H>  ENTR SUPR SUPR SUPR SUPR SUPR SUPR SUPR SUPR SUPR SUPR SUPR SUPR SUPR SUPR
 3H>  ENTR SUPR SUPR SUPR SUPR SUPR SUPR SUPR SUPR SUPR SUPR SUPR SUPR SUPR SUPR
 4H>  ENTR SUPR SUPR SUPR SUPR SUPR SUPR SUPR SUPR SUPR SUPR SUPR SUPR SUPR SUPR
 5H>  ENTR SUPR SUPR SUPR SUPR SUPR SUPR SUPR SUPR SUPR SUPR SUPR SUPR SUPR SUPR
 6H>  ENTR SUPR SUPR SUPR SUPR SUPR SUPR SUPR SUPR SUPR SUPR SUPR SUPR SUPR SUPR
 7H>  ENTR SUPR SUPR SUPR SUPR SUPR SUPR SUPR SUPR SUPR SUPR SUPR SUPR ANNU ANNU
 8H>  ENTR SUPR SUPR SUPR SUPR SUPR SUPR SUPR SUPR SUPR SUPR SUPR ANNU ANNU ANNU
 9H>  ENTR SUPR SUPR SUPR SUPR SUPR SUPR SUPR SUPR SUPR SUPR ANNU ANNU ANNU ANNU
10H>  ENTR SUPR SUPR SUPR SUPR SUPR SUPR SUPR SUPR SUPR ANNU ANNU ANNU ANNU ANNU
11H>  ENTR ANNU ANNU ANNU ANNU ANNU ANNU ANNU ANNU ANNU ANNU ANNU ANNU ANNU ANNU
12H>  ENTR ANNU ANNU ANNU ANNU ANNU ANNU ANNU ANNU ANNU ANNU ANNU ANNU ANNU ANNU
13H>  ENTR ANNU ANNU ANNU ANNU ANNU ANNU ANNU ANNU ANNU ANNU ANNU ANNU ANNU ANNU
14H>  ENTR ANNU ANNU ANNU ANNU ANNU ANNU ANNU ANNU ANNU ANNU ANNU ANNU ANNU ANNU
15H>  ENTR ANNU ANNU ANNU ANNU ANNU ANNU ANNU ANNU ANNU ANNU ANNU ANNU ANNU ANNU

16H<  ANNU ANNU ANNU ANNU ANNU ANNU ANNU ANNU ANNU ANNU ANNU ANNU ANNU ANNU BEND
17H<  ANNU ANNU ANNU ANNU ANNU ANNU ANNU ANNU ANNU ANNU ANNU ANNU ANNU ANNU BEND
18H<  ANNU ANNU ANNU ANNU ANNU ANNU ANNU ANNU ANNU ANNU ANNU ANNU ANNU ANNU BEND
19H<  ANNU ANNU ANNU ANNU ANNU ANNU ANNU ANNU ANNU ANNU ANNU ANNU ANNU ANNU BEND
20H<  ANNU ANNU ANNU ANNU ANNU ANNU ANNU ANNU ANNU ANNU ANNU ANNU ANNU ANNU BEND
21H<  ANNU ANNU ANNU ANNU ANNU ANNU ANNU ANNU ANNU ANNU ANNU ANNU ANNU ANNU BEND
22H<  ANNU ANNU ANNU ANNU ANNU ANNU ANNU ANNU ANNU ANNU ANNU ANNU ANNU ANNU BEND
23H<  ANNU ANNU ANNU ANNU ANNU ANNU ANNU ANNU ANNU ANNU ANNU ANNU ANNU ANNU BEND
24H<  ANNU ANNU ANNU ANNU ANNU ANNU ANNU ANNU ANNU ANNU ANNU ANNU ANNU ANNU BEND
25H<  ANNU ANNU ANNU ANNU ANNU ANNU ANNU ANNU ANNU ANNU ANNU ANNU ANNU ANNU BEND
26H<  ANNU ANNU ANNU ANNU ANNU ANNU ANNU ANNU ANNU ANNU ANNU ANNU ANNU ANNU BEND
27H<  ANNU ANNU ANNU ANNU ANNU ANNU ANNU ANNU ANNU ANNU ANNU ANNU ANNU ANNU BEND
28H<  ANNU ANNU ANNU ANNU ANNU ANNU ANNU ANNU ANNU ANNU ANNU ANNU ANNU ANNU BEND
29H<  ANNU ANNU ANNU ANNU ANNU ANNU ANNU ANNU ANNU ANNU ANNU ANNU ANNU ANNU BEND
30H<  SLUG SLUG ANNU ANNU ANNU ANNU ANNU ANNU ANNU ANNU ANNU ANNU ANNU ANNU BEND

31L<  ANNU ANNU ANNU ANNU ANNU ANNU ANNU ANNU ANNU ANNU ANNU ANNU ANNU ANNU ENTR
32L<  ANNU ANNU ANNU ANNU ANNU ANNU ANNU ANNU ANNU ANNU ANNU ANNU ANNU ANNU ENTR
33L<  ANNU ANNU ANNU ANNU ANNU ANNU ANNU ANNU ANNU ANNU ANNU ANNU ANNU ANNU ENTR
34L<  SUBC SUBC SLUG SLUG ANNU ANNU ANNU ANNU ANNU ANNU ANNU ANNU ANNU ANNU ENTR
35L<  SUBC SUBC SLUG SLUG ANNU ANNU ANNU ANNU ANNU ANNU ANNU ANNU ANNU ANNU ENTR
36L<  SUBC SUBC SLUG SLUG ANNU ANNU ANNU ANNU ANNU ANNU ANNU ANNU ANNU ANNU ENTR
37L<  SUBC SUBC SUBC SLUG SLUG ANNU ANNU ANNU ANNU ANNU ANNU ANNU ANNU ANNU ENTR
38L<  SUBC SUBC SUBC SLUG SLUG WAVY ANNU ANNU ANNU ANNU ANNU ANNU ANNU ANNU ENTR
39L<  SUBC SUBC SUBC SUBC SUBC SLUG SLUG ANNU ANNU ANNU ANNU ANNU ANNU ANNU ENTR
40L<  SUBC SUBC SUBC SUBC SUBC SUBC SUBC SLUG WAVY ANNU ANNU ANNU ANNU ANNU ENTR

41L>  BEND SUBC SUBC SUBC SUBC SUBC SUBC SUBC SUBC SUBC SUBC SUBC SUBC SUBC SUBC
42L>  BEND SUBC SUBC SUBC SUBC SUBC SUBC SUBC SUBC SUBC SUBC SUBC SUBC SUBC SUBC
43L>  BEND SUBC SUBC SUBC SUBC SUBC SUBC SUBC SUBC SUBC SUBC SUBC SUBC SUBC SUBC
44L>  BEND SUBC SUBC SUBC SUBC SUBC SUBC SUBC SUBC SUBC SUBC SUBC SUBC SUBC SUBC
45L>  BEND SUBC SUBC SUBC SUBC SUBC SUBC SUBC SUBC SUBC SUBC SUBC SUBC SUBC SUBC

46L<  SUBC SUBC SUBC SUBC SUBC SUBC SUBC SUBC SUBC SUBC SUBC SUBC SUBC SUBC ENTR
47L<  SUBC SUBC SUBC SUBC SUBC SUBC SUBC SUBC SUBC SUBC SUBC SUBC SUBC SUBC ENTR
48L<  SUBC SUBC SUBC SUBC SUBC SUBC SUBC SUBC SUBC SUBC SUBC SUBC SUBC SUBC ENTR
49L<  SUBC SUBC SUBC SUBC SUBC SUBC SUBC SUBC SUBC SUBC SUBC SUBC SUBC SUBC ENTR
50L<  SUBC SUBC SUBC SUBC SUBC SUBC SUBC SUBC SUBC SUBC SUBC SUBC SUBC SUBC ENTR

51L>  BEND SUBC SUBC SUBC SUBC SUBC SUBC SUBC SUBC SUBC SUBC SUBC SUBC SUBC SUBC
52L>  BEND SUBC SUBC SUBC SUBC SUBC SUBC SUBC SUBC SUBC SUBC SUBC SUBC SUBC SUBC
53L>  BEND SUBC SUBC SUBC SUBC SUBC SUBC SUBC SUBC SUBC SUBC SUBC SUBC SUBC SUBC
54L>  BEND SUBC SUBC SUBC SUBC SUBC SUBC SUBC SUBC SUBC SUBC SUBC SUBC SUBC SUBC
55L>  BEND SUBC SUBC SUBC SUBC SUBC SUBC SUBC SUBC SUBC SUBC SUBC SUBC SUBC SUBC
56L>  BEND SUBC SUBC SUBC SUBC SUBC SUBC SUBC SUBC SUBC SUBC SUBC SUBC SUBC SUBC
57L>  BEND SUBC SUBC SUBC SUBC SUBC SUBC SUBC SUBC SUBC SUBC SUBC SUBC SUBC SUBC
58L>  BEND ANNU SLUG SUBC SUBC SUBC SUBC SUBC SUBC SUBC SUBC SUBC SUBC SUBC SUBC
59L>  BEND ANNU ANNU ANNU SLUG SUBC SUBC SUBC SUBC SUBC SUBC SUBC SUBC SUBC SUBC
60L>  BEND ANNU ANNU ANNU SLUG SLUG SUBC SUBC SUBC SUBC SUBC SUBC SUBC SUBC SUBC
```

The fist column contains a number (tube row) plus a direction (< or >). Much of the top section is superheated (i.e., vapor only) and much of the bottom section is sub-cooled (i.e., liquid only). This is very undesirable, as the huge temperature differences along the tubes (superheated to sub-cooled) causes unequal thermal expansion and mechanical failure. The inputs in this case (including an orifice plate with differing diameter holes for each pipe) are based on actual operation, not hypothetical conditions. This next figure shows the temperature distribution throughout the tubes (blue=370°F to red=680°F):

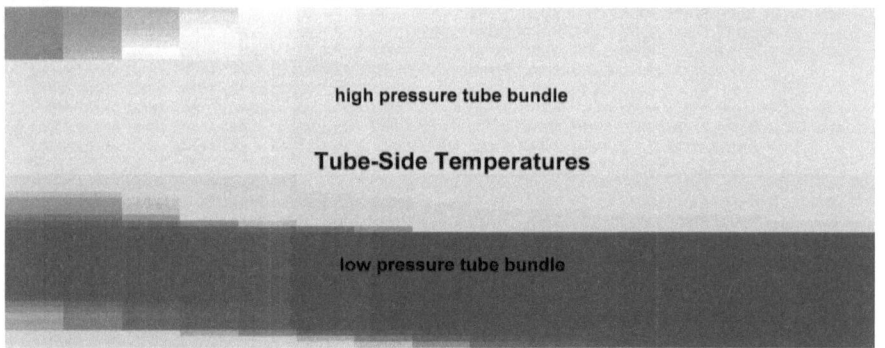

The distribution throughout the shell is similar:

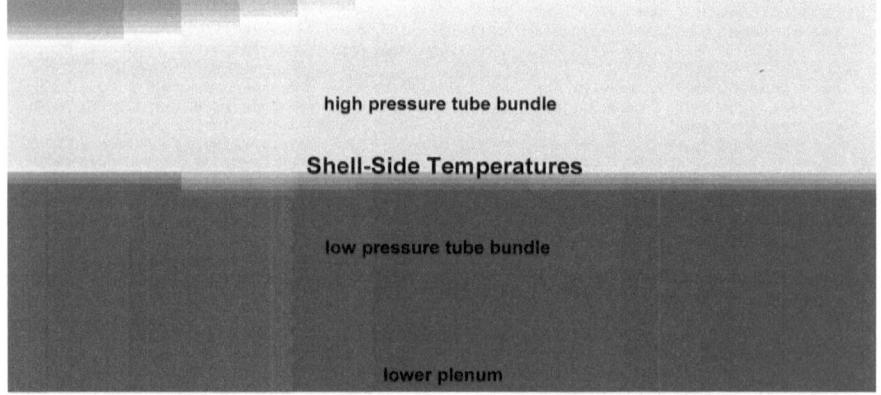

The void fraction is an important parameter in two-phase flow. A value of one indicates complete only vapor (saturated or superheated) and a value of zero indicates only liquid (saturated or sub-cooled). These correspond to the flow regimes shown in the previous figure, that is 1.000 corresponds to SUPR, 0.000 corresponds to SUBC, and everything in between corresponds to ANNUlar, WAVY, SLUG, etc. The predicted compressive stress in kilo-pounds per square inch is shown in the second figure.

VOID FRACTION (VAPOR AREA/TOTAL AREA)

Row														
1H>	0.000	1.000	1.000	1.000	1.000	1.000	1.000	1.000	1.000	1.000	1.000	1.000	1.000	1.000
2H>	0.000	1.000	1.000	1.000	1.000	1.000	1.000	1.000	1.000	1.000	1.000	1.000	1.000	1.000
3H>	0.000	1.000	1.000	1.000	1.000	1.000	1.000	1.000	1.000	1.000	1.000	1.000	1.000	1.000
4H>	0.000	1.000	1.000	1.000	1.000	1.000	1.000	1.000	1.000	1.000	1.000	1.000	1.000	1.000
5H>	0.000	1.000	1.000	1.000	1.000	1.000	1.000	1.000	1.000	1.000	1.000	1.000	1.000	1.000
6H>	0.000	1.000	1.000	1.000	1.000	1.000	1.000	1.000	1.000	1.000	1.000	1.000	1.000	1.000
7H>	0.000	1.000	1.000	1.000	1.000	1.000	1.000	1.000	1.000	1.000	1.000	1.000	1.000	0.997	0.997
8H>	0.000	1.000	1.000	1.000	1.000	1.000	1.000	1.000	1.000	1.000	1.000	1.000	0.991	0.974	0.974
9H>	0.000	1.000	1.000	1.000	1.000	1.000	1.000	1.000	1.000	1.000	1.000	0.977	0.957	0.957
10H>	0.000	1.000	1.000	1.000	1.000	1.000	1.000	1.000	1.000	1.000	0.968	0.936	0.911	0.911
11H>	0.000	0.409	0.407	0.405	0.404	0.403	0.401	0.400	0.399	0.398	0.398	0.397	0.396	0.396	0.396
12H>	0.000	0.408	0.406	0.403	0.402	0.400	0.398	0.397	0.396	0.395	0.394	0.393	0.392	0.392	0.392
13H>	0.000	0.407	0.404	0.401	0.399	0.396	0.394	0.392	0.391	0.389	0.388	0.387	0.387	0.386	0.386	0.386
14H>	0.000	0.406	0.401	0.397	0.394	0.391	0.388	0.386	0.384	0.382	0.380	0.379	0.379	0.378	0.378	0.378
15H>	0.000	0.404	0.398	0.393	0.388	0.384	0.380	0.377	0.374	0.372	0.370	0.369	0.368	0.368	0.368	0.368
16H<	0.373	0.373	0.374	0.374	0.374	0.375	0.375	0.375	0.376	0.376	0.376	0.377	0.377	0.377	0.378	0.000
17H<	0.383	0.383	0.384	0.384	0.385	0.385	0.385	0.386	0.386	0.386	0.387	0.387	0.388	0.388	0.000
18H<	0.391	0.391	0.391	0.392	0.392	0.393	0.393	0.393	0.394	0.394	0.394	0.395	0.395	0.395	0.396	0.000
19H<	0.396	0.396	0.397	0.397	0.398	0.398	0.398	0.399	0.399	0.399	0.400	0.400	0.401	0.401	0.401	0.000
20H<	0.400	0.400	0.401	0.401	0.402	0.402	0.402	0.403	0.403	0.403	0.404	0.404	0.405	0.405	0.000
21H<	0.499	0.499	0.522	0.544	0.569	0.594	0.622	0.651	0.682	0.713	0.746	0.779	0.814	0.848	0.882	0.000
22H<	0.472	0.472	0.495	0.518	0.542	0.567	0.595	0.624	0.659	0.695	0.734	0.776	0.819	0.865	0.911	0.000
23H<	0.407	0.407	0.431	0.454	0.479	0.506	0.535	0.566	0.601	0.639	0.684	0.734	0.789	0.847	0.910	0.000
24H<	0.333	0.333	0.356	0.379	0.405	0.432	0.462	0.496	0.534	0.576	0.624	0.678	0.745	0.820	0.905	0.000
25H<	0.259	0.259	0.281	0.303	0.328	0.354	0.384	0.418	0.457	0.501	0.552	0.612	0.684	0.775	0.885	0.000
26H<	0.189	0.189	0.208	0.228	0.251	0.275	0.304	0.336	0.375	0.419	0.471	0.535	0.613	0.711	0.848	0.000
27H<	0.131	0.131	0.147	0.165	0.185	0.207	0.233	0.263	0.299	0.342	0.394	0.459	0.542	0.651	0.804	0.000
28H<	0.080	0.080	0.094	0.108	0.126	0.144	0.167	0.194	0.226	0.265	0.314	0.377	0.460	0.575	0.746	0.000
29H<	0.041	0.041	0.052	0.064	0.078	0.094	0.113	0.136	0.164	0.198	0.243	0.301	0.381	0.498	0.683	0.000
30H<	0.008	0.008	0.017	0.028	0.039	0.052	0.068	0.087	0.111	0.141	0.180	0.234	0.310	0.427	0.629	0.000
31L<	0.050	0.050	0.058	0.067	0.076	0.085	0.096	0.109	0.126	0.144	0.168	0.198	0.237	0.287	0.355	0.000
32L<	0.028	0.028	0.035	0.042	0.050	0.057	0.067	0.079	0.094	0.111	0.133	0.162	0.201	0.253	0.331	0.000
33L<	0.011	0.011	0.017	0.023	0.029	0.036	0.044	0.054	0.067	0.082	0.103	0.130	0.167	0.221	0.305	0.000
34L<	0.000	0.000	0.001	0.006	0.011	0.017	0.023	0.032	0.043	0.056	0.074	0.098	0.132	0.185	0.274	0.000
35L<	0.000	0.000	0.000	0.004	0.007	0.014	0.020	0.028	0.039	0.052	0.069	0.092	0.125	0.176	0.264	0.000
36L<	0.000	0.000	0.001	0.004	0.008	0.014	0.021	0.029	0.039	0.052	0.069	0.092	0.125	0.176	0.265	0.000
37L<	0.000	0.000	0.000	0.002	0.006	0.012	0.019	0.027	0.037	0.050	0.067	0.089	0.122	0.173	0.261	0.000
38L<	0.000	0.000	0.000	0.000	0.004	0.008	0.015	0.022	0.032	0.044	0.060	0.082	0.114	0.163	0.252	0.000
39L<	0.000	0.000	0.000	0.000	0.000	0.001	0.005	0.013	0.021	0.032	0.047	0.067	0.097	0.144	0.233	0.000
40L<	0.000	0.000	0.000	0.000	0.000	0.000	0.000	0.002	0.013	0.025	0.042	0.067	0.109	0.193	0.000
41L>	0.000	0.000	0.000	0.000	0.000	0.000	0.000	0.000	0.000	0.000	0.000	0.000	0.000	0.000	0.000
42L>	0.000	0.000	0.000	0.000	0.000	0.000	0.000	0.000	0.000	0.000	0.000	0.000	0.000	0.000	0.000
43L>	0.000	0.000	0.000	0.000	0.000	0.000	0.000	0.000	0.000	0.000	0.000	0.000	0.000	0.000	0.000
44L>	0.000	0.000	0.000	0.000	0.000	0.000	0.000	0.000	0.000	0.000	0.000	0.000	0.000	0.000	0.000
45L>	0.000	0.000	0.000	0.000	0.000	0.000	0.000	0.000	0.000	0.000	0.000	0.000	0.000	0.000	0.000
46L<	0.000	0.000	0.000	0.000	0.000	0.000	0.000	0.000	0.000	0.000	0.000	0.000	0.000	0.000	0.000
47L<	0.000	0.000	0.000	0.000	0.000	0.000	0.000	0.000	0.000	0.000	0.000	0.000	0.000	0.000	0.000
48L<	0.000	0.000	0.000	0.000	0.000	0.000	0.000	0.000	0.000	0.000	0.000	0.000	0.000	0.000	0.000
49L<	0.000	0.000	0.000	0.000	0.000	0.000	0.000	0.000	0.000	0.000	0.000	0.000	0.000	0.000	0.000
50L<	0.000	0.000	0.000	0.000	0.000	0.000	0.000	0.000	0.000	0.000	0.000	0.000	0.000	0.000	0.000
51L>	0.000	0.000	0.000	0.000	0.000	0.000	0.000	0.000	0.000	0.000	0.000	0.000	0.000	0.000	0.000
52L>	0.000	0.000	0.000	0.000	0.000	0.000	0.000	0.000	0.000	0.000	0.000	0.000	0.000	0.000	0.000
53L>	0.000	0.000	0.000	0.000	0.000	0.000	0.000	0.000	0.000	0.000	0.000	0.000	0.000	0.000	0.000
54L>	0.000	0.000	0.000	0.000	0.000	0.000	0.000	0.000	0.000	0.000	0.000	0.000	0.000	0.000	0.000
55L>	0.000	0.000	0.000	0.000	0.000	0.000	0.000	0.000	0.000	0.000	0.000	0.000	0.000	0.000	0.000
56L>	0.000	0.000	0.000	0.000	0.000	0.000	0.000	0.000	0.000	0.000	0.000	0.000	0.000	0.000	0.000
57L>	0.000	0.000	0.000	0.000	0.000	0.000	0.000	0.000	0.000	0.000	0.000	0.000	0.000	0.000	0.000
58L>	0.000	0.009	0.003	0.000	0.000	0.000	0.000	0.000	0.000	0.000	0.000	0.000	0.000	0.000	0.000
59L>	0.000	0.024	0.016	0.009	0.003	0.000	0.000	0.000	0.000	0.000	0.000	0.000	0.000	0.000	0.000
60L>	0.000	0.042	0.030	0.020	0.013	0.006	0.002	0.000	0.000	0.000	0.000	0.000	0.000	0.000	0.000

COMPRESSIVE STRESS ON TUBES DUE TO THERMAL EXPANSION ASSUMING CLAMPED ENDS AND RIGID TUBE SHEETS

1H>	0	105.1	101.6	98.5	95.7	93.2	91.1	89.2	87.6	86.3	85.2	84.3	83.5	83.0	82.5	0
2H>	0	104.6	101.0	97.7	94.8	92.4	90.2	88.5	87.0	85.7	84.7	83.8	83.1	82.6	82.2	0
3H>	0	104.1	100.2	96.9	94.0	91.5	89.4	87.7	86.3	85.1	84.1	83.4	82.8	82.3	82.0	0
4H>	0	103.5	99.5	96.0	93.1	90.6	88.6	87.0	85.6	84.5	83.6	82.9	82.4	82.0	81.7	0
5H>	0	102.9	98.6	95.1	92.1	89.7	87.8	86.2	85.0	84.0	83.2	82.5	82.1	81.8	81.5	0
6H>	0	102.3	97.7	94.1	91.2	88.8	87.0	85.5	84.3	83.4	82.7	82.2	81.8	81.5	81.3	0
7H>	0	101.6	96.8	93.1	90.1	87.9	86.1	84.8	83.7	82.9	82.3	81.8	81.5	81.3	81.0	0
8H>	0	100.7	95.7	91.9	89.1	86.9	85.3	84.0	83.1	82.4	81.9	81.5	81.2	80.9	80.9	0
9H>	0	99.8	94.5	90.7	87.9	85.8	84.3	83.3	82.5	81.9	81.5	81.2	81.0	80.8	80.8	0
10H>	0	98.8	93.0	89.1	86.4	84.6	83.3	82.4	81.8	81.4	81.1	80.9	80.7	80.7	80.7	0
11H>	0	80.7	80.7	80.7	80.7	80.6	80.6	80.6	80.5	80.5	80.5	80.4	80.4	80.4	80.4	0
12H>	0	80.7	80.7	80.7	80.6	80.6	80.6	80.6	80.5	80.5	80.5	80.4	80.4	80.4	80.4	0
13H>	0	80.7	80.6	80.6	80.6	80.6	80.5	80.5	80.5	80.5	80.5	80.4	80.4	80.4	80.4	0
14H>	0	80.6	80.6	80.6	80.6	80.5	80.5	80.5	80.5	80.5	80.4	80.4	80.4	80.4	80.4	0
15H>	0	80.5	80.5	80.5	80.5	80.5	80.4	80.4	80.4	80.4	80.4	80.4	80.4	80.4	80.4	0
16H<	0	79.9	79.9	80.0	80.0	80.0	80.1	80.1	80.1	80.2	80.2	80.3	80.3	80.3	80.4	0
17H<	0	79.9	79.9	80.0	80.0	80.0	80.1	80.1	80.1	80.2	80.2	80.3	80.3	80.3	80.4	0
18H<	0	79.9	79.9	80.0	80.0	80.0	80.1	80.1	80.1	80.2	80.2	80.3	80.3	80.3	80.4	0
19H<	0	79.9	79.9	80.0	80.0	80.0	80.1	80.1	80.1	80.2	80.2	80.3	80.3	80.3	80.4	0
20H<	0	79.9	79.9	80.0	80.0	80.0	80.1	80.1	80.1	80.2	80.2	80.3	80.3	80.3	80.4	0
21H<	0	80.4	80.4	80.4	80.4	80.4	80.5	80.5	80.5	80.5	80.5	80.6	80.6	80.6	80.6	0
22H<	0	80.3	80.3	80.4	80.4	80.4	80.4	80.3	80.4	80.4	80.4	80.5	80.5	80.5	80.6	0
23H<	0	80.1	80.1	80.2	80.2	80.2	80.2	80.2	80.3	80.2	80.2	80.3	80.3	80.3	80.4	0
24H<	0	79.8	79.9	80.0	80.0	80.0	80.0	80.1	80.1	80.1	80.0	80.1	80.1	80.1	80.2	0
25H<	0	79.5	79.5	79.7	79.7	79.7	79.7	79.7	79.8	79.8	79.8	79.7	79.8	79.9	0	
26H<	0	79.1	79.1	79.3	79.3	79.3	79.3	79.3	79.4	79.4	79.4	79.3	79.4	0		
27H<	0	78.5	78.5	78.7	78.7	78.7	78.7	78.7	78.8	78.8	78.9	78.9	78.9	78.8	0	
28H<	0	77.9	77.9	78.1	78.1	78.1	78.0	78.0	78.1	78.1	78.2	78.2	78.2	78.2	78.0	0
29H<	0	77.2	77.2	77.4	77.4	77.3	77.2	77.1	77.1	77.2	77.2	77.2	77.2	77.3	76.9	0
30H<	0	75.8	75.7	76.5	76.6	76.5	76.2	76.2	76.1	76.1	76.1	76.1	76.1	76.1	76.0	0
31L>	0	64.5	64.5	65.3	65.3	65.2	65.0	65.0	65.0	65.1	65.2	65.2	65.3	65.3	65.4	0
32L>	0	63.9	63.9	64.8	64.7	64.6	64.3	64.3	64.3	64.5	64.5	64.5	64.6	64.7	64.7	0
33L>	0	63.0	63.2	64.2	64.1	63.9	63.5	63.5	63.4	63.6	63.7	63.7	63.7	63.8	63.8	0
34L>	0	62.6	64.5	62.8	63.3	63.2	62.6	62.5	62.4	62.4	62.4	62.4	62.4	62.5	62.5	0
35L>	0	60.4	64.2	62.0	62.3	62.7	62.6	62.5	62.3	62.2	62.2	62.1	62.0	62.0	61.9	0
36L<	0	60.2	63.5	61.9	62.3	62.7	62.6	62.4	62.3	62.2	62.1	62.1	62.0	62.0	61.9	0
37L>	0	59.9	62.9	63.8	62.0	62.6	62.6	62.5	62.3	62.2	62.2	62.1	62.0	62.0	61.9	0
38L>	0	58.3	60.6	64.6	63.6	62.4	62.8	62.6	62.5	62.4	62.3	62.2	62.1	62.0	62.0	0
39L>	0	56.2	57.5	59.6	63.0	64.6	63.7	62.9	62.7	62.5	62.4	62.3	62.2	62.1	0	
40L>	0	54.6	54.9	55.5	56.4	58.1	60.9	65.0	64.5	63.3	62.9	63.0	62.8	62.7	62.5	0
41L>	0	54.2	54.2	54.2	54.2	54.2	54.2	54.2	54.2	54.2	54.2	54.2	54.2	54.2	54.2	0
42L>	0	54.2	54.2	54.2	54.2	54.2	54.2	54.2	54.2	54.2	54.2	54.2	54.2	54.2	54.2	0
43L>	0	54.2	54.2	54.2	54.2	54.2	54.2	54.2	54.2	54.2	54.2	54.2	54.2	54.2	54.2	0
44L>	0	54.2	54.2	54.2	54.2	54.2	54.2	54.2	54.2	54.2	54.2	54.2	54.2	54.2	54.2	0
45L>	0	54.2	54.2	54.2	54.2	54.2	54.2	54.2	54.2	54.2	54.2	54.2	54.2	54.2	54.2	0
46L<	0	54.2	54.2	54.2	54.2	54.2	54.2	54.2	54.2	54.2	54.2	54.2	54.2	54.2	54.3	0
47L<	0	54.2	54.2	54.2	54.2	54.2	54.2	54.2	54.2	54.2	54.2	54.2	54.2	54.2	54.3	0
48L<	0	54.2	54.2	54.2	54.2	54.2	54.2	54.2	54.2	54.2	54.2	54.2	54.2	54.2	54.3	0
49L<	0	54.2	54.2	54.2	54.2	54.2	54.2	54.2	54.2	54.2	54.2	54.2	54.2	54.2	54.3	0
50L<	0	54.2	54.2	54.2	54.2	54.2	54.2	54.2	54.2	54.2	54.2	54.2	54.2	54.2	54.3	0
51L>	0	54.4	54.3	54.3	54.2	54.2	54.2	54.2	54.2	54.2	54.2	54.2	54.2	54.2	54.2	0
52L>	0	55.2	54.8	54.6	54.5	54.4	54.3	54.3	54.3	54.2	54.2	54.2	54.2	54.2	54.2	0
53L>	0	56.9	56.0	55.3	54.9	54.7	54.5	54.4	54.3	54.3	54.3	54.2	54.2	54.2	54.2	0
54L>	0	57.9	56.6	55.8	55.2	54.9	54.6	54.5	54.4	54.3	54.3	54.3	54.3	54.2	54.2	0
55L>	0	59.1	57.4	56.3	55.6	55.1	54.8	54.6	54.5	54.4	54.3	54.3	54.3	54.3	54.2	0
56L>	0	58.6	57.1	56.1	55.4	55.0	54.7	54.6	54.4	54.4	54.3	54.3	54.3	54.3	54.2	0
57L>	0	58.8	57.2	56.2	55.5	55.0	54.8	54.4	54.4	54.4	54.3	54.3	54.3	54.3	54.2	0
58L>	0	62.5	63.8	61.8	59.1	57.4	56.2	55.5	55.1	54.8	54.6	54.5	54.4	54.3	54.3	0
59L>	0	62.7	62.9	62.6	63.8	62.2	59.3	57.5	56.3	55.5	55.1	54.8	54.6	54.4	54.4	0
60L>	0	62.5	62.6	62.8	62.8	62.1	63.4	60.2	58.0	56.6	55.7	55.2	54.8	54.6	54.5	0

Chapter 15. Condensation in Crossflow

In this chapter we will consider the condensation of steam on the outside of tubes in a crossflow when the velocity is non-trivial. Such is definitely the case when steam exiting a low-pressure turbine may reach sonic velocity, turns downward in a hood, and impinges on the top row of tubes in a large water-cooled condenser. By the time the steam reaches the bottom row of tubes (that isn't flooded), the kinetic energy is spent and the velocity is negligible. This interesting process occurs because, at typical condenser operating conditions (1.5 in.HgA), the density of water in the liquid state is 25,000 times that of the corresponding vapor.

The original experimental work on this type of condensation and associated formula for calculating the heat transfer coefficient is due to Nusselt. This formula may be found in any heat transfer text as well as readily on-line. Shekriladze and Gomelauri modified Nusselt's correlation[21] to account for non-trivial vapor velocity.

$$Nu = 0.728 \, \text{Re}^{\frac{1}{2}} = \frac{hd}{k} \qquad (15.1)$$

where the film Reynolds number is given by:

$$\text{Re} = \left[\frac{\rho_f (\rho_f - \rho_g) h_{fg} g d^3}{k \mu \Delta T} \right]^{\frac{1}{2}} \qquad (15.2)$$

Thermal resistance to heat transfer in the condensation of steam on a cooled surface arises from the thin film of liquid that forms and must run off the surface as condensation proceeds. A crossflow in the vapor produces a shear stress, which strips away the liquid, thinning the film and reducing the resistance. This is the basis for Shekriladze and Gomelauri modification.[22] The modified Nusselt number is given by:

$$Nu = \frac{0.9 + 0.728 F^{\frac{1}{2}}}{\left(1 + 3.44 F^{\frac{1}{2}} + F^2\right)^{\frac{1}{4}}} \, \text{Re}^{\frac{1}{2}} \qquad (15.3)$$

---

[21] Nusselt, W., "Die Oberflächenkondensation des Wasserdampfes [The Surface Condensation of Water Vapor]," VDI [Association of German Engineers], 1916.

[22] Shekriladze, I. G. and Gomelauri, V. I., "Theoretical Study of Laminar Film Condensation of Flowing Vapor," International Journal of Heat and Mass Transfer, Vol. 9, pp. 581–591, 1966.

F is equal to the Prandtl number divided by Froude and Jacob number, or:

$$F = \left(\frac{\mu h_{fg}}{uk\Delta T}\right)\sqrt{\frac{g}{d}} \qquad (15.4)$$

This formulation can be applied to several types of heat exchangers, including a water-cooled steam surface condenser such as would be found in a power plant. The inputs to such a model include geometry and operating conditions. Thermal and transport properties are also needed. These are provided on the *properties* tab and are accessed via an interpolation macro, also included in the spreadsheet condensation_crossflow.xls.

	A	B	C	D
1	Condensation on Horizontal Tubes in Crossflow			
2	INPUTS	symbol	units	value
3	Steam Turbine Exhaust			
4	turbine annulus area	Aan	ft²	72
5	number of ends	Nan	-	4
6	Condenser Geometry			
7	tube length	L	ft	40.0
8	number of tubes	N	-	12,000
9	tube outside dia.	do	in	1.125
10	tube gauge	ga	-	22
11	tube wall conduct.	kw	BTU/hr/ft/°F	9.32
12	tube pitch	pitch	in	1.125
13	Steam			
14	steam flow	stm	lbm/hr	2,500,000
15	steam quality	x	-	91%
16	Cooling Water			
17	flow	Qccw	gpm	240,000
18	inlet temperature	Tccw,in	°F	60
19	SOLUTION			
20	backpressure	Psat	in.HgA	1.84

The solution is the operating pressure (or turbine backpressure). This is solved iteratively using a bisection search when you push the button provided. The cooling water inlet temperature is an input and the outlet temperature is calculated from the duty (i.e., the steam condensed).

21	PROPERTIES	symbol	units	value
22	Cooling Water			
23	density	ρ	lbm/ft³	62.37
24	specific heat	Cp	BTU/lbm/°F	1.0004
25	viscosity	μ	lbm/ft/hr	2.712
26	thermal cond.	k	BTU/hr/ft/°F	0.3423
27	Prandtl Number	Pr	-	7.93
28	Steam			
29	liquid density	ρf	lbm/ft³	62.01
30	vapor density	ρg	lbm/ft³	0.0027
31	latent heat	hfg	BTU/lbm	1037.6
32	liquid viscosity	μf	lbm/ft/hr	1.678
33	liquid thermal cond.	kf	BTU/hr/ft/°F	0.3611
34	CALCULATIONS	symbol	units	value
35	Condenser Geometry			
36	tube wall thick.	wt	in	0.028
37	tube inside dia.	di	in	1.069
38	water flow area	Af	ft²	74.79
39	surface area	As	ft²	141,372
40	steam flow area	Ap	ft²	410.79

The film Reynolds number depends weakly on the temperature difference across the condensing film. A (spatially) constant value is used here, equal to the overall log-mean temperature difference.

41	Cooling Water			
42	duty	duty	BTU/hr	2.358E+09
43	mass flow rate	mccw	lbm/hr	1.201E+08
44	outlet temperature	Tccw,out	°F	79.63
45	velocity inside tubes	ut	ft/sec	7.15
46	tube side Reynolds	Re,t	-	52,729
47	tube side Nusselt	Nu,t	-	315
48	tube side ht. tr. coef.	ht	BTU/hr/ft/°F	1212
49	tube wall ht. tr. coef.	hw	BTU/hr/ft/°F	3894
50	Steam			
51	operating pressure	Psat	psia	0.9046
52	saturation temp.	Tsat	°F	98.38
53	exit specific volume	1/ρan	ft³/lbm	332.5
54	annulus velocity	Van	ft/sec	802
55	log mean temp. diff.	LMTD	°F	27.4
56	required conductance	U	BTU/hr/°F	608.9
57	film Reynolds number	Re	-	2.874E+05

The heat transfer for each row of tubes (in 20 groups) is calculated sequentially so that the local steam velocity can be used to calculate the impact on condensing heat transfer coefficient.

Local Heat Transfer Calculations								
vertical position	ht. tr. BTU/hr	quality -	sp.vol. ft³/lbm	velocity ft/sec	F -	Nu -	hc BTU/hr/ft²/°F	U BTU/hr/ft²/°F
top	1.21E+08	86.23%	315.4	533.1	6.12	553.2	2130	627.2
2	1.21E+08	81.55%	298.3	504.2	6.47	550.4	2120	626.3
3	1.21E+08	76.88%	281.2	475.4	6.86	547.5	2108	625.3
4	1.21E+08	72.22%	264.2	446.6	7.30	544.3	2096	624.2
5	1.21E+08	67.57%	247.1	417.8	7.80	540.8	2083	623.0
6	1.20E+08	62.93%	230.2	389.1	8.38	537.1	2069	621.8
7	1.20E+08	58.30%	213.2	360.5	9.05	533.1	2053	620.3
8	1.20E+08	53.68%	196.3	331.9	9.82	528.8	2036	618.8
9	1.20E+08	49.07%	179.5	303.4	10.75	524.1	2018	617.1
10	1.19E+08	44.48%	162.7	275.0	11.85	518.9	1999	615.3
11	1.19E+08	39.90%	146.0	246.7	13.21	513.4	1977	613.2
12	1.18E+08	35.34%	129.3	218.5	14.92	507.2	1954	610.9
13	1.18E+08	30.80%	112.7	190.5	17.12	500.5	1928	608.4
14	1.17E+08	26.28%	96.1	162.5	20.06	493.0	1899	605.5
15	1.17E+08	21.79%	79.7	134.7	24.20	484.7	1867	602.2
16	1.16E+08	17.32%	63.4	107.1	30.44	475.2	1830	598.3
17	1.15E+08	12.89%	47.1	79.7	40.91	464.1	1788	593.7
18	1.14E+08	8.50%	31.1	52.6	62.02	450.7	1736	587.9
19	1.12E+08	4.17%	15.3	25.8	126.34	432.9	1667	579.8
bottom	1.08E+08	0.00%	0.0	0.0	∞	390.3	1503	558.6

The variation in Nusselt number is illustrated in the following figure:

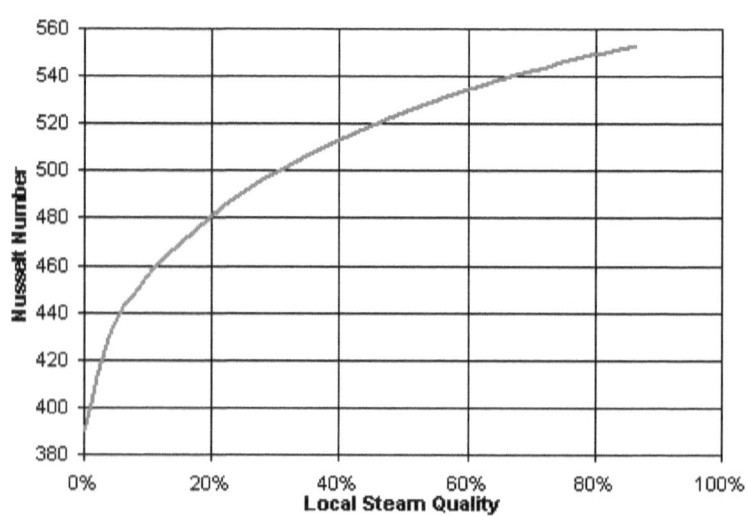

78

There is also a button and macro to vary the cooling water inlet temperature and produce a table of backpressures:

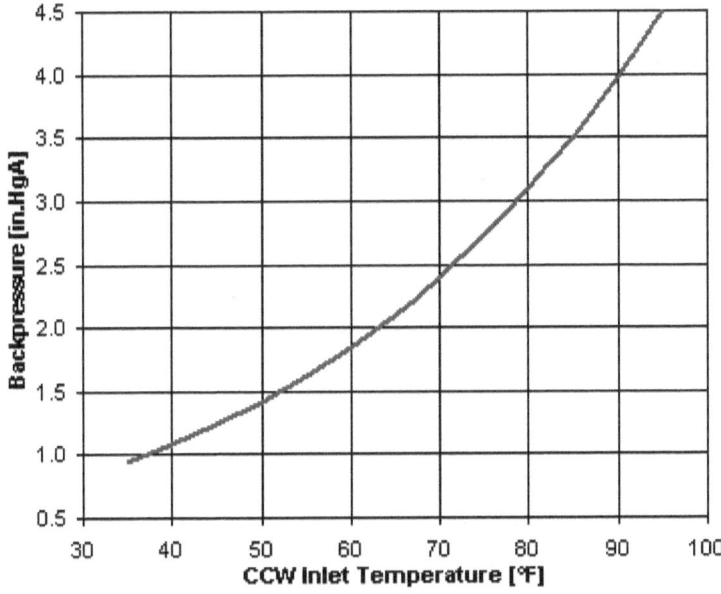

Chapter 16. Operational Data

It is often assumed that all data exhibit normally distributed variability; however, this is often not the case. We will first consider some 7966 hourly averages representing the performance of an actual high-pressure feedwater heater. The total heat transfer and overall conductance (following the analysis in Chapter 7) are shown in this first figure as it varies throughout the year.

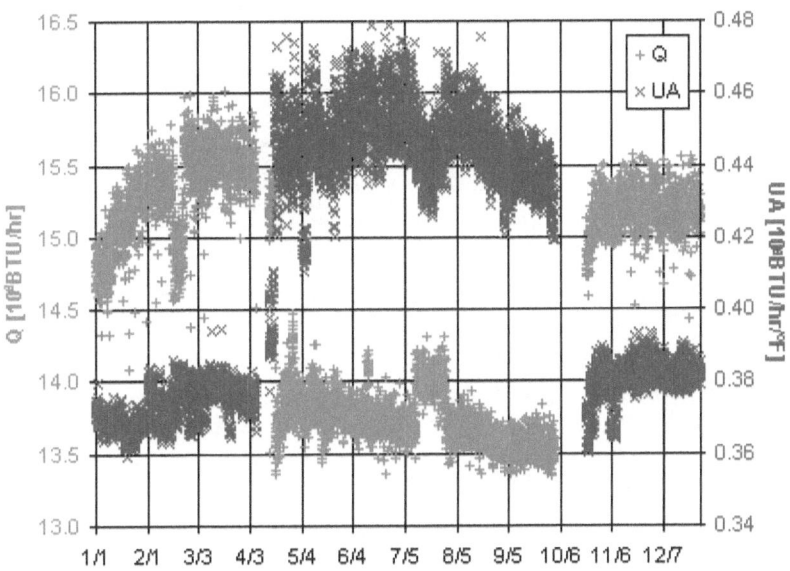

There are clearly two modes of operation: one in the colder months and another in the warmer months. This arises from a change in boiler operation to meet seasonal emissions targets. The probability of occurrence of this same data is shown in the second figure.

The two modes of operation produce two peaks in the probability. Note that the area under the red (Q) and blue (UA) curves both sum to unity (i.e., 100%) so that the horizontal scale is immaterial. The split between these two modes of operation is so close that the area under each peak sums to one-half (i.e., 50%). The difference is only 0.25%. Such distinctions don't occur by chance. This is a response to a purposefully implemented standard operating procedure (SOP).

This data can be presented in statistical form by counting the number of occurrences in each interval over the range of observed values. The resulting probability distributions are fairly normal, as is often the case with actual operating data of this sort. The variability or spread of the data so plotted is an indication of the uncertainty of the measurements as well as the operation in addition to the two distinct modes. All of the data and graphs are in the spreadsheet HP_FWH_operational.xls, which is included in the on-line archive.

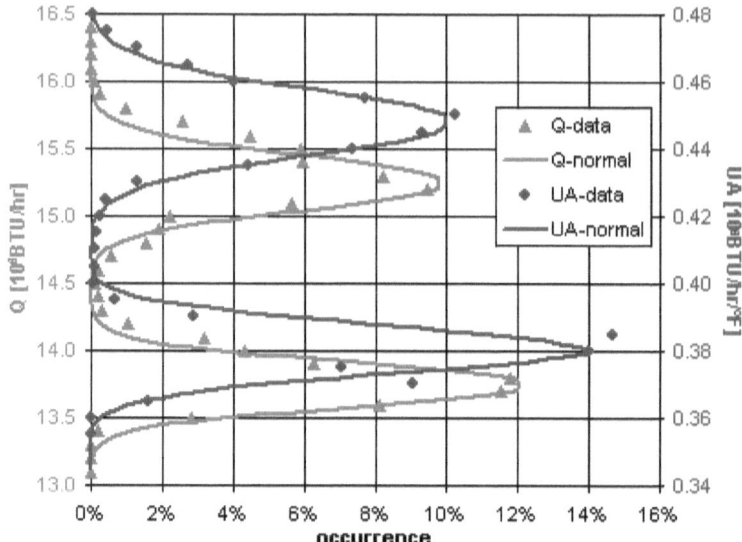

The heat transfer, Q, is well approximated by two combined normal distributions: 13.75±0.16 and 15.25±0.20 million BTU/hr. The conductance, UA, is also well approximated by two combined normal distributions: 0.3803±0.0071 and 0.4480±0.0098 million BTU/hr/°F.

We will next consider similar operational data, this time from a condenser. As the water flow in this case is based on pump curves, rather than an accurate measurement, this cannot be considered a code-level test in accordance with PTC-12.5; nevertheless, this is often the best that can be done under the circumstances. At this plant, the condenser cooling water (CCW) flows through buried concrete pipe with no accessible straight run in support of a pitot traverse, the preferred method of measurement. Dye dilution was also rejected for lack of an acceptable injection point. The dimensions of the condenser are:

DESIGN INFORMATION		
description	units	value
number of shells (HP/LP)	-	2
area (per shell)	ft²	195,000
nominal steam flow (per shell)	lbm/hr	1,665,000
nominal duty (per shell)	BTU/hr	1.20E+09
number of CCW pumps	-	3
nominal CCW flow	gpm	111,267
TESTING INFORMATION		
condensate nozzle diameter	in	7.500
condensate nozzle Cd	-	0.9021

This is the beginning of the operational data:

	A	B	C	D	E	F	G	H	I	J
1					MEASURED DATA					
2	Description	CCW inlet temp.	CCW exit temp.	CCW pump head	LPTA exh. pres. gov. end	LPTA exh. pres. gen. end	LPTB exh. pres. gov. end	LPTB exh. pres. gen. end	hotwell temp.	condensate flow nozzle delta-P
3	Units	Deg F	Deg F	in.H2O	in.HgA	in.HgA	in.HgA	in.HgA	Deg F	in.H2O
4	2/26/14 9:27	80.29	93.54	186.39	1.701	1.718	2.105	2.077	100.95	408.23
5	2/26/14 9:27	80.28	93.53	188.35	1.707	1.722	2.100	2.074	100.94	408.79
6	2/26/14 9:28	80.28	93.53	185.39	1.703	1.718	2.099	2.072	100.94	407.92
7	2/26/14 9:28	80.26	93.51	192.77	1.703	1.721	2.096	2.069	100.91	408.72
8	2/26/14 9:29	80.25	93.49	187.39	1.709	1.726	2.104	2.076	100.89	408.67
9	2/26/14 9:29	80.23	93.48	187.86	1.711	1.726	2.106	2.077	100.89	405.63
10	2/26/14 9:30	80.23	93.47	186.17	1.713	1.730	2.112	2.084	100.90	402.02
11	2/26/14 9:30	80.24	93.48	189.36	1.716	1.733	2.116	2.087	100.95	401.51
12	2/26/14 9:31	80.22	93.46	184.60	1.715	1.728	2.113	2.086	101.00	407.29
13	2/26/14 9:31	80.22	93.46	186.40	1.716	1.735	2.111	2.085	101.03	415.64
14	2/26/14 9:32	80.22	93.46	186.64	1.720	1.735	2.111	2.085	101.04	413.42
15	2/26/14 9:32	80.22	93.46	187.58	1.725	1.740	2.120	2.094	101.05	404.64
16	2/26/14 9:33	80.21	93.44	182.51	1.727	1.742	2.127	2.100	101.11	401.63
17	2/26/14 9:33	80.21	93.45	184.38	1.732	1.749	2.137	2.109	101.14	402.46
18	2/26/14 9:34	80.20	93.43	186.15	1.736	1.751	2.138	2.107	101.22	405.96
19	2/26/14 9:34	80.19	93.43	185.74	1.736	1.754	2.142	2.115	101.29	409.87
20	2/26/14 9:35	80.20	93.43	182.34	1.742	1.755	2.143	2.116	101.37	409.32
21	2/26/14 9:35	80.20	93.43	185.42	1.740	1.759	2.146	2.118	101.44	404.79

This is the calculation section:

K	L	M	N	O	P	Q	R	S	T	U	V	W	X
						CALCULATIONS							
CCW flow	duty	cond. density	cond. flow	LPTA exh. pres.	LPTB exh. pres.	LPTA exh. temp.	LPTB exh. temp.	LPTA exhaust quality	LPTB exhaust quality	shell A LMTD	shell B LMTD	shell A U	shell B U
gpm	BTU/hr	lbm/ft³	lbm/hr	in.HgA	in.HgA	Deg F	Deg F	-	-	Deg F	Deg F	BTU/hr/ft²/°F	
457,219	3.018E+09	61.980	3,372,124	1.710	2.091	95.88	102.54	0.8612	0.8644	11.98	12.02	645.9	644.0
457,129	3.017E+09	61.980	3,374,446	1.715	2.087	95.98	102.48	0.8604	0.8635	12.08	11.95	640.3	647.1
457,268	3.018E+09	61.980	3,370,850	1.711	2.086	95.90	102.46	0.8615	0.8646	12.01	11.94	644.2	648.1
457,140	3.016E+09	61.980	3,374,185	1.712	2.082	95.93	102.41	0.8603	0.8634	12.06	11.91	641.5	649.6
457,148	3.016E+09	61.980	3,373,994	1.717	2.090	96.04	102.53	0.8604	0.8635	12.18	12.04	635.1	642.1
457,635	3.019E+09	61.981	3,361,398	1.718	2.091	96.05	102.55	0.8645	0.8676	12.21	12.08	633.9	640.6
458,209	3.023E+09	61.980	3,346,434	1.722	2.098	96.11	102.66	0.8694	0.8725	12.28	12.20	631.3	635.4
458,291	3.023E+09	61.980	3,344,267	1.725	2.101	96.17	102.71	0.8700	0.8732	12.32	12.24	629.5	633.1
457,369	3.017E+09	61.979	3,368,202	1.722	2.099	96.11	102.68	0.8620	0.8651	12.28	12.23	629.7	632.4
456,025	3.007E+09	61.979	3,402,518	1.725	2.098	96.19	102.66	0.8508	0.8538	12.37	12.21	623.6	631.6
456,384	3.010E+09	61.979	3,393,416	1.728	2.098	96.22	102.66	0.8536	0.8567	12.40	12.21	622.3	632.1
457,793	3.019E+09	61.979	3,357,203	1.733	2.107	96.32	102.80	0.8655	0.8686	12.50	12.36	619.2	626.3
458,271	3.021E+09	61.978	3,344,665	1.734	2.113	96.36	102.90	0.8696	0.8727	12.55	12.48	617.3	620.9

The same '67 steam property functions used in the spreadsheet. Perhaps the most critical measurement is the turbine exhaust pressure. In this case there are two low-pressure turbines, each with two ends: one on the governor end and the other on the generator end. The average pressures are shown in this next figure:

83

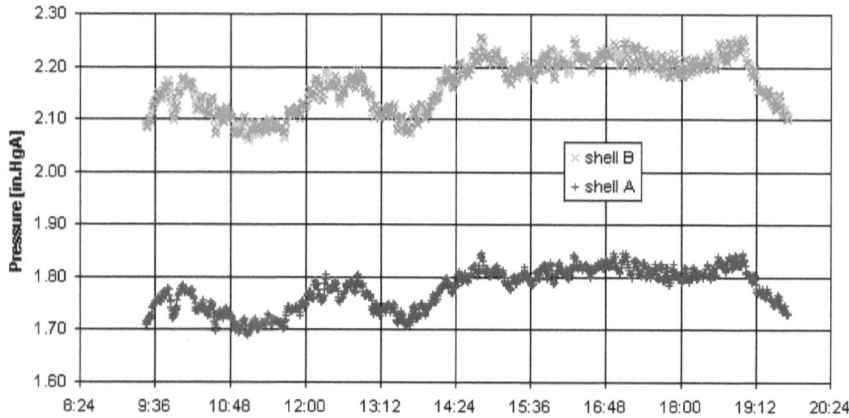

The saturation temperature associated with all four pressures is consistent with the hotwell temperature, something of great concern in these tests. If the hotwell temperature is higher, then the pressures cannot possibly be correct. If the hotwell temperature is more than 1.8°F/1.0°C lower (i.e., *sub-cooling*), the pressures are suspect. Here, this is not an issue.

Condenser cooling water flow was determined by measuring the head across the pumps and the curves. This is rarely an accurate determination of flow. If it were, there would be no point in testing pumps. Such curves are often based on scale models and may easily be off by 15%. In this case, the indicated flow was remarkably close to the design.

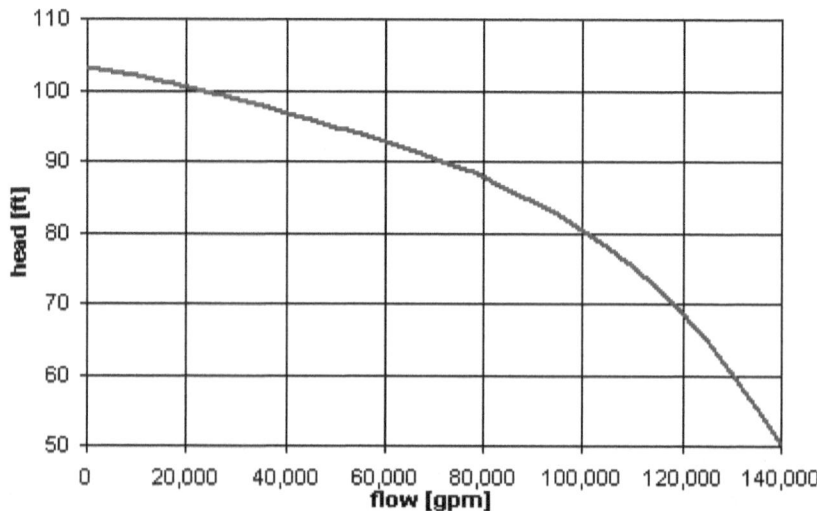

The LMTDs are well behaved, stable, and also close to the design values:

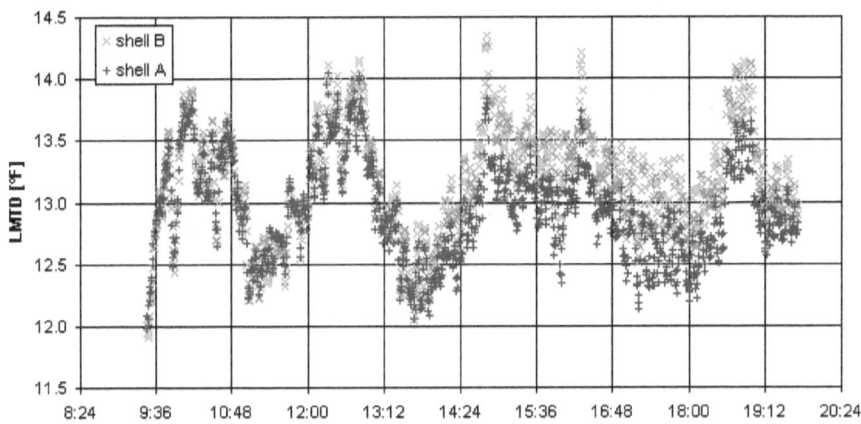

The two sections of the condenser are side-by-side. Although there were two fittings where temperature probes were inserted, there is no reason to presume that these two points accurately represent the average temperature between the two sections. The duty of the two sections was assumed equal in these calculations for lack of any better option. The overall conductance, U, of each shell is shown in this next figure.

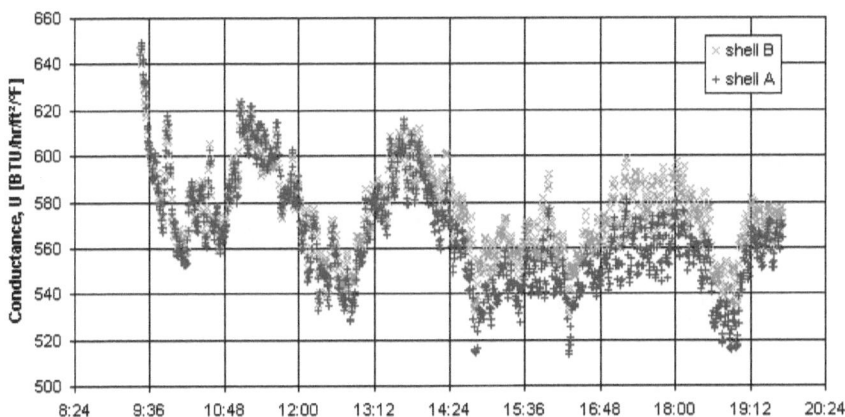

This data indicates a single mode of operation and all of the variables exhibited a fairly normal distribution of values. Only the U is shown here, as this was the result of greatest interest and a direct indication of the cleanliness of the condenser.

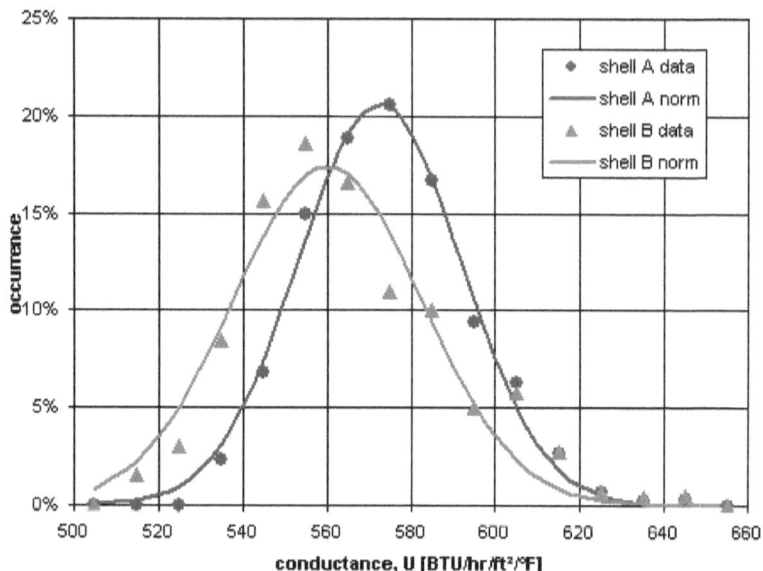

The A shell (blue dots) has a mean of 572.6 with a standard deviation of 19.28 BTU/hr/ft²/°F and the B shell (red triangles) has a mean of 560.2 with a standard deviation of 22.3 BTU/hr/ft²/°F.

Chapter 17. Monte Carlo Methods

A model or computational method utilizing random values is often called Monte Carlo after the famous casino. Monte Carlo models have several uses in heat exchanger analysis, including the estimate of uncertainty in calculated results (e.g., heat transfer, mean temperature difference, and overall conductance) based on the uncertainty of the inputs (e.g., flows, temperatures, and properties). In this chapter we will explore this method of arriving at uncertainty.

At the core of any Monte Carlo method is a function to generate normally distributed random numbers. Most random number generators (e.g., the rand() function in Excel® and it's equivalent in VBA® rnd()), return uniformly distributed random numbers. What this means is that, if you create a large list of such numbers, sort these, separate them into groups, and count the number in each group, you will get a nearly flat curve. That's not what we want. For a Monte Carlo model, we must have numbers that form a bell-shaped curve.

The simplest way of generating normally distributed random numbers from uniformly distributed ones is to take the average of several of the latter. Experience has shown that 12 are sufficient to accomplish the desired result. Sample code is provided in Appendix C.

The following spreadsheet illustrates the Monte Carlo Method for estimating heat exchanger uncertainty:

	A	B	C	D	E	F	G	H	I
1	Monte Carlo Simulation (10,000 cases)								
2	name	units	mean	uncer					
3	mH	kg/s	4.00	0.40	push to update				
4	CpH	kJ/kg/°C	3.50	0.05					
5	mC	kg/s	9.00	0.90					
6	CpC	kJ/kg/°C	2.50	0.04					
7	THi	°C	500	0.67					
8	THo	°C	300	0.67					
9	TCi	°C	250	0.50					
10	TCo	°C	375	0.50					
11	A	m²	100	N/A					
12									
13	case	mH	CpH	mC	CpC	THi	THo	TCi	TCo
14	0	4.000	3.500	9.000	2.500	500.0	300.0	250.0	375.0
15	1	4.025	3.496	9.275	2.503	499.9	300.3	250.1	374.9
16	2	3.901	3.494	8.911	2.493	500.1	300.1	250.0	374.8
17	3	3.937	3.487	9.114	2.498	499.6	300.0	250.0	374.9
18	4	3.895	3.505	8.636	2.485	500.2	299.9	250.0	375.0
19	5	4.135	3.502	9.077	2.504	499.8	300.1	250.1	374.8
20	6	3.886	3.508	8.889	2.508	500.0	299.7	250.1	375.0

While the randomized values could be entered into the spreadsheet with function calls, doing so results in a prohibitive recalculation time (>1 hour). Instead, a button is provided that fills in the cells with the same values that would be obtained by using function calls. As both flows and all four temperatures are measured, this results in an ambiguity (or discrepancy) between the hot and cold side heat transfer and overall conductance. The heat transfer for both sides is illustrated in this first figure, ΔT in the second, and U in the third:

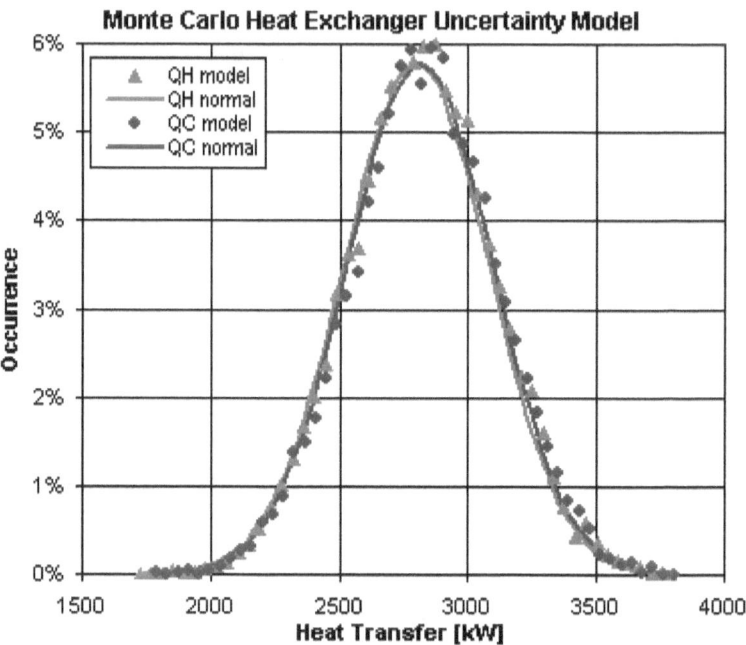

It should be no surprise that the results fall out on a bell-shaped curve. This is to be expected, considering the individual points were generated with normally distributed random numbers. The Monte Carlo Method quickly translates uncertainty in the inputs to uncertainty in the outputs. While it is possible to obtain this result analytically, the algebra is quite cumbersome. We might consider the following equation:

$$(\lambda_{Qh} \pm \sigma_{Qh}) = (\lambda_{\dot{m}h} \pm \sigma_{\dot{m}h})(\lambda_{Cph} \pm \sigma_{Cph})[(\lambda_{Thi} \pm \sigma_{Thi}) - (\lambda_{Tho} \pm \sigma_{Tho})] \quad (17.1)$$

where λ_X represents the mean of X and σ_X represents the uncertainty or variation of X. As there are four occurrences of $\pm$ on the right side of Equation 17.1, there are eight possible values, even with all of the parameters held constant. A similar situation exists for the LMTD:

$$(\lambda_{LMTD} \pm \sigma_{LMTD}) = \frac{(\lambda_{\Delta T1} \pm \sigma_{\Delta T1}) - (\lambda_{\Delta T2} \pm \sigma_{\Delta T2})}{\log\left(\dfrac{\lambda_{\Delta T1} \pm \sigma_{\Delta T1}}{\lambda_{\Delta T2} \pm \sigma_{\Delta T2}}\right)} \quad (17.2)$$

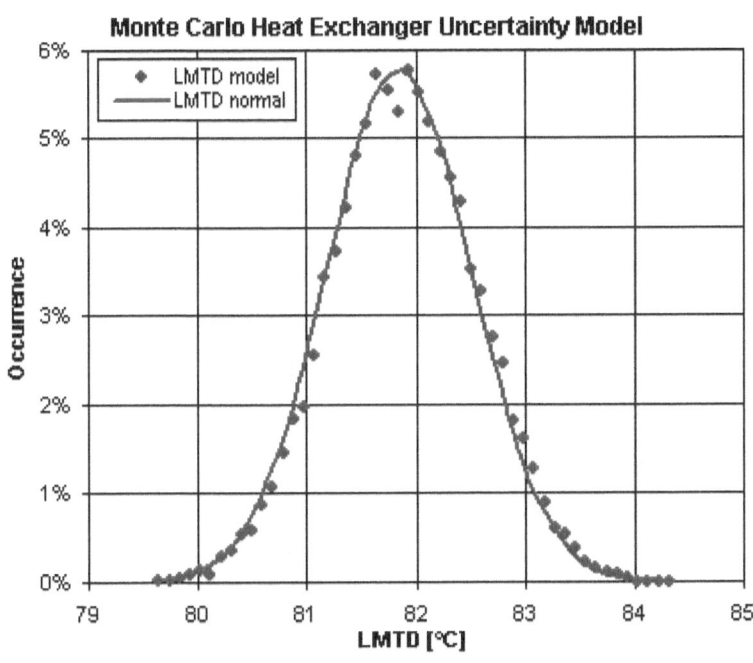
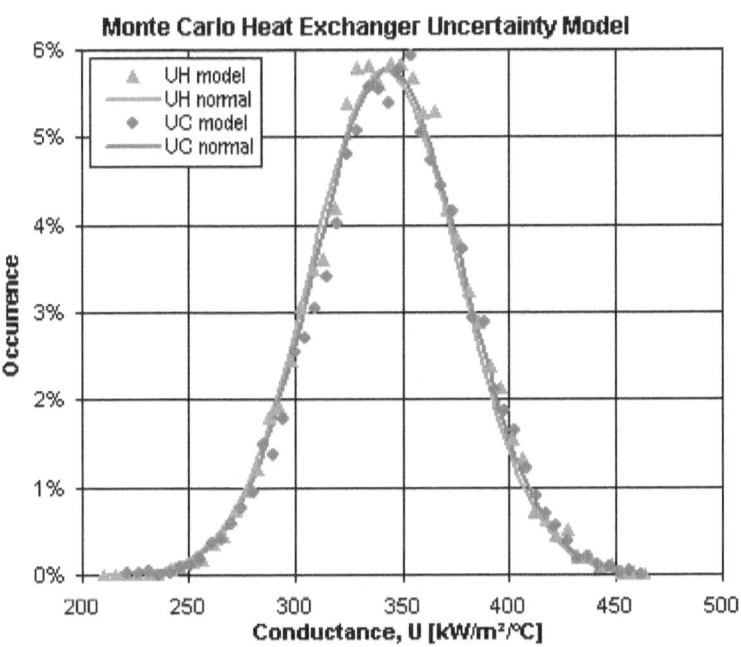

This simulation with 10,000 cases is adequate to achieve occurrence levels that well approximate bell-shaped curves, as illustrated in the previous three figures. Not surprisingly, the Monte Carlo simulation is much closer to the ideal than actual data in the previous chapter. Ten thousand is about the upper limit for Excel®. Real, compiled code can easily handle one million cases, which yields smooth bell-shaped curves, as shown in these next figures. The code is listed in Appendix C.

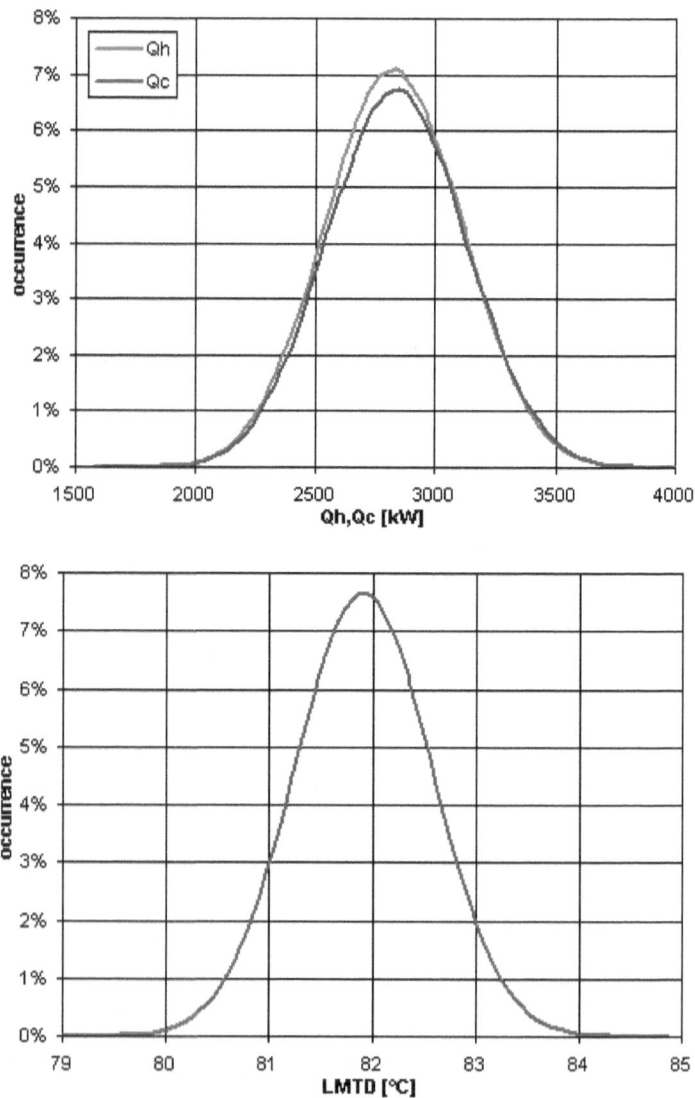

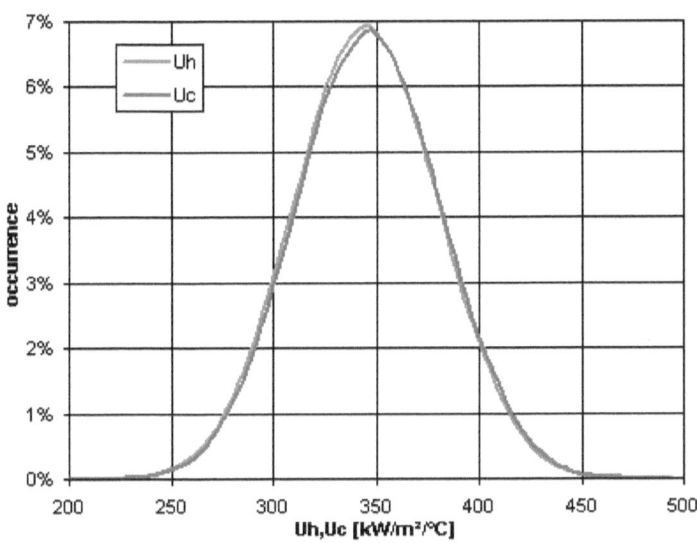

Appendix A. Crossflow Program

The following program implements the unmixed crossflow calculation discussed in Chapter 3 with and without iterative implicit temperature differences plus the closed-form implicit algorithm. Some sections containing trivial code have been omitted. The complete program is included in the on-line archive.

```
double CpH=4.0;
double CpC=1.0;
double mH=100.;
double mC=500.;
double Thi=50.;
double Tho=20.;
double Tci=15.;

double Ts(double Tn,double Tw,double UA,int nx,int ny)
{
  double alpha,beta,delta,gamma;
  beta=UA/mH/CpH;
  alpha=mC*CpC/mH/CpH;
  delta=2*alpha*nx*ny;
  gamma=(ny*ny+nx*nx*alpha*alpha+delta)*beta*beta
    +delta*delta;
  return(Tn+((Tn-Tw)*(ny+nx*alpha)*beta*beta-beta*(Tn-
    Tw)*sqrt(gamma))/delta/ny);
}

double solve(int nx,int ny,int m)
{
  int i,iter,x,y;
  double *dT,dT1,dT2,*dQ,*Tc,tco,*Th,tho,U1,U2,UA;

/* allocate arrays */

  dT=alloc(nx*ny,sizeof(double));
  dQ=alloc(nx*ny,sizeof(double));
  Th=alloc(nx*(ny+1),sizeof(double));
  Tc=alloc((nx+1)*ny,sizeof(double));

/* initialize arrays */

  for(x=0;x<nx;x++)
    Th[x]=Thi;
  for(y=0;y<ny;y++)
    Tc[(nx+1)*y]=Tci;

/* initialize UA */

  tco=Tci+mH*CpH*(Thi-Tho)/mC/CpC;
```

```
      U1=mH*CpH*(Thi-Tho)/LMTD(Thi-tco,Tho-Tci);
      U2=8.*U1;
      for(iter=0;iter<32;iter++)
        {
        UA=(U1+U2)/2.;

/* step through grid */

        for(y=0;y<ny;y++)
          {
          for(x=0;x<nx;x++)
            {
            if(m>=0)
              {
              dT[nx*y+x]=fmax(0.,Th[nx*y+x]-Tc[(nx+1)*y+x]);
              dQ[nx*y+x]=(UA/nx/ny)*dT[nx*y+x];
              Th[nx*(y+1)+x]=Th[nx*y+x]-dQ[nx*y+x]/
    CpH/(mH/nx);
              }
            else
              {
              Th[nx*(y+1)+x]=Ts(Th[nx*y+x],Tc[(nx+1)*y+x],
    UA,nx,ny);
              dQ[nx*y+x]=(Th[nx*y+x]-Th[nx*(y+1)+x])
    *CpH*(mH/nx);
              dT[nx*y+x]=dQ[nx*y+x]/(UA/nx/ny);
              }
            Tc[(nx+1)*y+x+1]=Tc[(nx+1)*y+x]+dQ[nx*y+x]
    /CpC/(mC/ny);
            }
          }

/* implicit temperature correction */

        for(i=0;i<m;i++)
          {
          for(y=0;y<ny;y++)
            {
            for(x=0;x<nx;x++)
              {
              dT1=Th[nx*y+x]-Tc[(nx+1)*y+x];
              dT2=Th[nx*(y+1)+x]-Tc[(nx+1)*y+x+1];
              dT[nx*y+x]=fmax(0.,(2.*dT1+dT2)/3.);
              dQ[nx*y+x]=(UA/nx/ny)*dT[nx*y+x];
              Th[nx*(y+1)+x]=Th[nx*y+x]-
    dQ[nx*y+x]/CpH/(mH/nx);
              Tc[(nx+1)*y+x+1]=Tc[(nx+1)*y+x]+dQ[nx*y+x]/
    CpC/(mC/ny);
              }
```

```
        }
      }
/* solve for exit temperatures */

   for(tho=x=0;x<nx;x++)
     tho+=Th[nx*ny+x];
   tho/=nx;
   for(tco=y=0;y<ny;y++)
     tco+=Tc[(nx+1)*y+nx];
   tco/=ny;

/* use bisection search algorithm, adjusting UA to match
   Tho */

   if(tho>Tho)
     U1=UA;
   else
     U2=UA;
   }

/* release memory */

  free(dT);
  free(dQ);
  free(Th);
  free(Tc);

  return(UA);
  }

void test4(int m)
  {
  int n;
  for(n=5;n<=100;n++)
    printf("%i %lG\n",n,solve(n,n,m));
  }

int main(int argc,char**argv,char**envp)
  {
  test4(-1);
  return(0);
  }
```

The following is a sample of the output:

```
nx  ny    UA
10  10  1724.26
10  20  1941.27
10  30  2024.87
10  40  2069.15
20  10  1887.29
20  20  2150.29
20  30  2253.31
20  40  2308.28
30  10  1948.66
30  20  2230.29
30  30  2341.32
30  40  2400.72
40  10  1980.85
40  20  2272.56
40  30  2387.94
40  40  2449.75
```

Appendix B. Moisture Separator/Reheater Program

The analysis of two-phase flow inside tubes presented in Chapter 14 is only a part of the code necessary to model a moisture separator/reheater. The entire code consists of several sections, some of which will be listed here. The entire code, along with a pre-compiled Windows® executable, is included in the on-line archive.

The program, called MISER, first checks the inputs, then establishes the tube connections, then initializes the variables, then solves the equations, iterating until convergence is reached. The two main sections of interest are the two-phase flow calculations within a computational element of tube and the combined tube-side/shell-side calculation.

Tube-Side Module

```
/************************************************************
    Determine the two-phase flow frictional pressure
       gradient
    using Chisholm's method for separated flow
    (this method is reportedly valid for steam above 435
       psia but it
    is about the best two-phase pressure drop correlation
       available
    so it will be used anyway...)

    Determine the two-phase heat transfer coefficient for
       horizontal
    pipe flow using  the Taitel-Dukler method

    For more deatils see "Prediction of Horizontal
       Tubeside
    Condensation of Pure Components Using Flow Regime
       Criteria,"
    G. Breber, J. W. Palen, and J. Taborek, 18th ASME
       National
    Heat Transfer Conference, San Diego, 1979.

    a........ tube cross sectional area [sq.ft.]
    c........ constant (see Collier equation 2.72, p.50)
    c2....... constant (see Collier equation 2.72, p.50)
    cbar..... constant (see Collier equation 2.75, p.51)
    cpf...... constant pressure specific heat of saturated
       liquid
    cpg...... constant pressure specific heat of saturated
       vapor
    d........ tube diameter [ft]
    dpdx..... two-phase frictional pressure gradient
```

dpdxf.... frictional pressure gradient based on liquid alone
dpdxg.... frictional pressure gradient based on vapor alone
dt....... the temperature difference across the condensing vapor
ff....... friction factor - liquid alone
fg....... friction factor - vapor alone
fm....... friction factor - mixture
flow..... flow [#/hr]
fp....... two-phase flow pattern (ie. annuar, wavy, slug etc.)
fr....... Froude number
g........ gravitational acceleration (32.174 [ft/sec**2])
gm....... mass flux [#/sq.ft./hr]
gf....... mass flux of liquid flowing alone [#/sq.ft./hr]
gg....... mass flux of vapor flowing alone [#/sq.ft./hr]
gstar.... critical mass flux [#/sq.ft./hr] (see Collier p.50)
gsubc.... Newton's constant (32.174 [lbm-ft/lbf/sec**2])
hf....... enthaply of saturated liquid
hg....... enthaply of saturated vapor
h2p...... two-phase heat transfer coefficient [BTU/hr/f/sq.ft.]
tf....... thermal condictivity of the saturated liquid [BTU/hr/f/ft]
tg....... thermal condictivity of the saturated vapor [BTU/hr/f/ft]
gamma.... constant (see Collier equation 7.72, p.50)
b........ Blasius exponent (see Collier p.50)
dvf...... viscosity of saturated liquid [#/ft/hr]
dvg...... viscosity of saturated vapor [#/ft/hr]
dvm...... viscosity of mixture [#/ft/hr]
phif..... two phase multiplier (see Collier equation 2.50, p.34)
phif2.... square of the two phase multiplier (phif**2)
prf...... Prandtl number of the saturated liquid
psi...... two phase multiplier (see Collier equation 2.73, p.51)
ref...... Reynolds number for liquid flow only
reg...... Reynolds number for vapor flow only
rem...... Reynolds number of mixture
slip..... slip ratio (vapor velocity/liquid velocity)
x........ quality

```
    xx....... Lockart-Martinelli factor (see Collier
      equation 2.67, p. 37)
    ug....... linear momentum flux [#/ft/hr**2]
    vf....... specific volume of saturated liquid
      [cu.ft./#]
    vg....... specific volume of saturated vapor
      [cu.ft./#]
    alpha.... void fraction (cross-sectional area of vapor
      flow/
            cross-sectional area of total flow)
****************************************************/

void phase2(double *p,double *x,double *flow,double
    *a,double *d,double *rem,double *fm,double
    *dpdx,double *dt,double *h2p,double *slip,double
    *alpha,double *ug,double *fp,int *ier)
    {
    int izone;
    double
      b,c,c2,cbar,cpf,cpg,dpdxf,dpdxg,dvf,dvg,dvm,f,ff,fg,f
      r,ftp,gamma,gf,gg,gm,gstar,hcg,hcs,hf,hfg,hg,hl,hv,ph
      if,phif2,prf,prg,psi,ref,reg,rhof,rhog,t,tf,tg,tsat,v
      ,vf,vfg,vg,xx;
    static BOOL rough=FALSE;
/* fetch thermodynamic and transport properties */
    tsat=fTsat(*p);
    vf=VfofT(tsat);
    vg=VgofT(tsat);
    hf=HfofT(tsat);
    hg=HgofT(tsat);
    dvf=MUfofT(tsat);
    dvg=MUgofT(tsat);
    tf=TKfofT(tsat);
    tg=TKgofT(tsat);
    cpf=CpfofT(tsat);
    cpg=CpgofT(tsat);
    vfg=vg-vf;
    rhof=1./vf;
    rhog=1./vg;
    hfg=hg-hf;
    prf=cpf*dvf/tf;
    prg=cpg*dvg/tg;
/* determine frictional pressure gradient for subcooled
    liquid */
    if((*x)<=0.0001)
        {
        gf=(*flow)/(*a);
        ref=gf*(*d)/dvf;
        *rem=ref;
```

```
            ff=fblas(ref,roughness);
            *fm=ff;
            *dpdx=-4.*ff*sq(gf/3600.)*vf/2./gsubc/(*d)/144.;
            *slip=0.;
            *alpha=0.;
            *ug=sq(gf)*vf;
            *fp=flopat(5);
      /* set two-phase heat transfer coefficient equal to the
         forced convection
         heat transfer coefficient (see Breber, Palen, and
         Taborek p.7) */
            *h2p=1.86*(tf/(*d))*pow(ref*prf,0.333333);
            if(ref>3000.)

            *h2p=0.024*(tf/(*d))*pow(ref,0.8)*pow(prf,0.333333);
      /* calculate the two-phase pressure drop for a
         liquid/vapor mixture (see Collier p.50) */
            }
          else if((*x)>0.9999)
            {
      /* determine frictional pressure gradient for
         superheated vapor */
            gg=*flow/(*a);
            reg=gg*(*d)/dvg;
            *rem=reg;
            fg=fblas(reg,roughness);
            *fm=fg;
            *dpdx=-4.*fg*sq(gg/3600.)*vg/2./gsubc/(*d)/144.;
            *slip=1.;
            *alpha=1.;
            *ug=gg*vg;
            *fp=flopat(6);
      /* set two-phase heat transfer coefficient equal to the
         forced convection
         heat transfer coefficient (see Breber, Palen, and
         Taborek p.7) */
            *h2p=1.86*(tg/(*d))*pow(reg*prg,0.333333);
            if(reg>3000.)

            *h2p=0.024*(tg/(*d))*pow(reg,0.8)*pow(prg,0.333333);
            }
          else
            {
            gamma=0.75;
            if(rough)
               gamma=1.;
            gstar=1.47E6;
            if(rough)
               gstar=1.1E6;
```

```
      b=0.;
      if(rough)
        b=0.25;
/* determine the effective viscosity (see Collier p.30)
   */
      dvm=1./((*x)/dvg+(1.-(*x))/dvf);
/* determine frictional pressure gradient for liquid
   alone */
      gf=*flow*(1.-(*x))/(*a);
      ref=gf*(*d)/dvf;
      ff=fblas(ref,roughness);
      dpdxf=-4.*ff*sq(gf/3600.)*vf/2./gsubc/(*d)/144.;
/* determine frictional pressure gradient for vapor
   alone */
      gg=(*flow)*(*x)/(*a);
      reg=gg*(*d)/dvg;
      fg=fblas(reg,roughness);
      dpdxg=-4.*fg*sq(gg/3600.)*vg/2./gsubc/(*d)/144.;
/* determine Lockhart-Martinelli parameter (see Collier
   p.37) */
      xx=sqrt(dpdxf/dpdxg);
/* (see Collier p.50) */
      gm=(*flow)/(*a);
      c2=fmin(2.*gamma,gstar/gm);
/* (see Collier p.51) */
      cbar=sqrt(vg/vf)+sqrt(vf/vg);
/* (see Collier p.50) */
      c=(gamma+(c2-gamma)*sqrt(vfg/vg))*cbar;
/* sub-critical flow (see Collier pages 50 & 51) */
      if(gm>gstar)
        {
/* super-critical flow (see Collier p.52) */
        t=pow((*x)/(1.-(*x)),(2.-
   b)/2.)*pow(dvf/dvg,b/2.)*sqrt(vf/vg);
        psi=(1.+c/t+1./sq(t))/(1.+cbar/t+1./sq(t));
        phif2=(1.+cbar/xx+1./sq(xx))*psi;
        }
      else
        phif2=1.+c/xx+1./sq(xx);
      *dpdx=dpdxf*phif2;
/* determine the effective friction factor and Reynolds
   number */
      gm=(*flow)/(*a);
      *rem=gm*(*d)/dvm;
      v=vf+(*x)*vfg;
      *fm=-(*dpdx)*144.*(*d)*gsubc*2./v/sq(gm/3600.)/4.;
/* calculate the two-phase multiplier */
      phif=sqrt(phif2);
```

```
/* calculate slip ratio using Zivi's method (ASME-JHT
   May, 1964 pages 247-252) */
   *slip=pow(vg/vf,0.333333);
/* calculate void fraction using zivi's method (ASME-JHT
   May, 1964, pages 247-252) */
   *alpha=1./(1.+((1.-(*x))/(*x))*sq(*slip));
/* determine the linear momentum flux */
   *ug=sq(gm)*(sq(*x)*vg/(*alpha)+sq(1.-(*x))*vf/(1.-
   (*alpha)));
/* Determine the two-phase flow regime and calculate the
   two-phase heat transfer coefficient. First determine
   the froude number (see Breber, Palen, and Taborek
   p.3) */
   fr=sqrt(sq(gg/3600.)/(*d)/g/rhog/(rhof-rhog));
/* determine the two-phase flow regime (see Breber,
   Palen, and Taborek p.7) */
   izone=1;
   if(xx<=1.25&&fr<=1.)
     izone=2;
   if(xx>1.25&&fr<=1.)
     izone=3;
   if(xx>1.25&&fr>1.)
     izone=4;
   *fp=flopat(izone);
/* calculate the liquid heat transfer coefficient
   (see Breber, Palen, and Taborek p.7) */
   hl=1.86*(tf/(*d))*pow(ref*prf,0.333333);
   if(ref>3000.)
     hl=0.024*(tf/(*d))*pow(ref,0.8)*pow(prf,0.333333);
/* calculate the heat transfer coefficient for
   condensing vapor (see Collier pages 328 & 331) */
   f=0.31*pow(reg,0.12);
   hv=f*pow(rhof*(rhof-
   rhog)*(g*3600.*3600.)*hfg*cube(tf)/(*d)/dvf/(*dt),0.2
   5);
/* determine the two-phase heat transfer coefficient
   (see Breber, Palen, and Taborek p.7) */
   ftp=pow(phif2,0.45);
   hcs=1./(1./(hl*ftp)+1./hv);
   fg=0.79;
   hcg=fg*hv;
/* heat transfer coefficient within zones i-iv
   (see Breber, Palen, and Taborek p.7) */
   if(izone==1)
     *h2p=hcs;
   if(izone==2)
     *h2p=hcg;
   if(izone==3)
     *h2p=hcs;
```

```
     if(izone==4)
        *h2p=hcs;
/* heat transfer coefficient between zone boundaries
   (see Breber, Palen, and Taborek p.7) */
     if(fr>.5&&fr<1.5)
        *h2p=hcs*(fr-0.5)+hcg*(1.5-fr);
     if(xx>.5&&xx<1.5)
        *h2p=hcs*(1.5-xx)+hcg*(xx-0.5);
    }
}
```

Shell-Side Module

```
void bundle(int is1,int is,int js,int it,int jt1,int
   jt,int ib)
{
int ibis;
double
   a0,ai,ao,at,bdt,cpg,dlt,dpdx,dpf,dpm,dt,dvg,gm,gmax,h
   ai,hao,haw,hc,hdk,hs1,ht1,pr,ps1,pt1,qmax,qmin,slip,t
   g,tsat,tubes,ua,ug1,us1,vs1,wall;
/* solve the integral equations for a cell in the tube
   bundles */
/* initialize new values using old values */
pt1=pt[it-1][jt1-1];
ht1=ht[it-1][jt1-1];
ug1=ug[it-1][jt1-1];
ps1=ps[is1-1][js-1];
vs1=vs[is1-1][js-1];
hs1=hs[is1-1][js-1];
us1=us[is1-1][js-1];
/* determine the heat exchange and flow areas */
tubes=(double)(ntr[it-1]-ntp[it-1]);
dlt=tl[ib-1]/(double)(NC-2);
if(jt==1||jt==NC)
   dlt=di[ib-1];
ai=M_PI*di[ib-1]*dlt*tubes;
a0=M_PI*d0[ib-1]*dlt*tubes;
ao=a0*af[ib-1];
at=M_PI*tubes*sq(di[ib-1])/4.;
/* determine the velocity */
ut[it-1][jt-1]=ft[it-1]*vt[it-1][jt1-1]/at/3600.;
if(rev[it-1])
   ut[it-1][jt-1]=-ut[it-1][jt-1];
/* determine the two-phase flow frictional pressure
   drop and heat transfer coefficient */
dt=fmax(1.,tt[it-1][jt1-1]-ts[is1-1][js-1]);
phase2(&pt[it-1][jt-1],&xt[it-1][jt-1],&ft[it-
   1],&at,&di[ib-1],&rt[it-1][jt-1],&fi[it-1][jt-
```

```
    1],&dpdx,&dt,&hi[it-1][jt-1],&slip,&al[it-1][jt-
    1],&ug[it-1][jt-1],&fp[it-1][jt-1],&ier);
  if(ier)
    error(ier);
  dpf=dpdx*dlt;
  /* determine the momentum pressure drop */
  dpm=(ug[it-1][jt-1]-ug1)/gsubc/2./3600./3600./144.;
  /* determine the pressure exiting the cell */
  pt[it-1][jt-1]=pt1+dpf+dpm;
  /* determine the heat transfer coefficient on the
     inside of the tubes */
  hai=hi[it-1][jt-1]*ai;
  /* determine the heat transfer coefficient for the
     tube wall including the scale (crud) on both sides */
  wall=2.*tk[ib-1]*(ao-ai)/log(ao/ai)/(d0[ib-1]-di[ib-
    1]);
  hc=750.;
  haw=1./(1./wall+1./(ao*hc));
  /* determine the maximum mass flux across the tubes in
     the shell */
  gm=fs[is-1][js-1]/dlt/wb[ib-1];
  gmax=gm*smin[ib-1]/(smin[ib-1]-d0[ib-1]);
  /* calculate the maximum local velocity in the shell
    */
  us[is-1][js-1]=gmax*vs[is-1][js-1]/3600.;
  /* compute thermodynamic and transport properties in
     the shell */
  tsat=fTsat(ps[is-1][js-1]);
  tg=TKgofT(tsat);
  cpg=CpgofT(tsat);
  dvg=MUgofT(tsat);
  pr=cpg*dvg/tg;
  /* calculate the frictional pressure drop by the
     method of Briggs and Young, Chem. Eng. Prog. Symp.
     Series No. 41, 59, 1965.
   * (see McGraw-Hill Handbook of Heat Transfer, p. 18-
     81) */
  rs[is-1][js-1]=gmax*dr[ib-1]/dvg;
  fo[is-1][js-1]=18.93*pow(st[ib-1]/smin[ib-
    1],0.515)/pow(rs[is-1][js-1],0.316)/pow(st[ib-
    1]/dr[ib-1],0.927);
  dpf=-fo[is-1][js-1]*sq(gmax/3600.)*vs[is-1][js-
    1]/gsubc/144.;
  /* calcualte the momentum pressure drop */
  dpm=(sq(us1)/vs1-sq(us[is-1][js-1])/vs[is-1][js-
    1])/2./gsubc/144.;
  /* calculate the pressure exiting the cell */
  ps[is-1][js-1]=ps1+dpf+dpm;
```

```
/* determine the heat transfer coefficient on the
outside of the tubes by the method of Robinson and
Briggs,8th ASME/AIChE
* National Heat Transfer Conference, No. 20,1965.
(see McGraw-Hill Handbook of Heat Transfer, p.18-81)
*/
hdk=0.134*pow(rs[is-1][js-
1],0.681)*pow(pr,0.333333)*pow(sp[ib-1]/fh[ib-
1],0.2)*pow(sp[ib-1]/fk[ib-1],0.113);
ho[it-1][jt-1]=hdk*tg/dr[ib-1];
hao=ho[it-1][jt-1]*ao;
/* determine the overall conductance "ua" */
ua=1./(1./hai+1./haw+1./hao);
if(jt==1||jt==NC)
  {
  hi[it-1][jt-1]=0.;
  ho[it-1][jt-1]=0.;
  ua=0.;
  fo[is-1][js-1]=0.;
  ps[is-1][js-1]=ps1;
  }
u0[it-1][jt-1]=ua/a0;
/* determine the temperature difference (using a
backward difference as this is conservative and never
violates the Second
* Law of Thermodynamics) and the heat transfer for
this cell */
qmin=0.;
qmax=fmax(0.,ua*(tt[it-1][jt1-1]-ts[is1-1][js-1]));
/* use bisection method to determine heat transfer */
for(ibis=1;ibis<=20;ibis++)
  {
  qt[it-1][jt-1]=(qmin+qmax)/2.;
  if(ibis>1&&qmax-qmin<=.001*qt[it-1][jt-1])
    break;
  /* determine the exiting enthalpy from the first law
  */
  ht[it-1][jt-1]=ht1-qt[it-1][jt-1]/ft[it-1];
  hs[is-1][js-1]=hs1+qt[it-1][jt-1]/fs[is-1][js-1];
  /* check for gross overshoot in the heat transfer if
  there is overshoot divide the heat transfer by two
  and go back to the
    * First Law */
  if(!(ht[it-1][jt-1]>1499.||hs[is-1][js-1] <
    1.))
    {
    if(!(ht[it-1][jt-1]<1.||hs[is-1][js-1] >
      1499.))
      {
```

```c
            /* determine the temperature,specific volume,and
      quality from the pressure and enthalpy */
            TofPH(pt[it-1][jt-1],ht[it-1][jt-1],&tt[it-
      1][jt-1],&vt[it-1][jt-1],&xt[it-1][jt-1]);
            TofPH(ps[is-1][js-1],hs[is-1][js-1],&ts[is-
      1][js-1],&vs[is-1][js-1],&xs[is-1][js-1]);
            /* bisection algorithm */
            if(!(qt[it-1][jt-1]>ua*(tt[it-1][jt-1]-ts[is-
      1][js-1])||tt[it-1][jt-1]<=ts[is-1][js-1]))
               goto L_100;
            }
         qmax=qt[it-1][jt-1];
         continue;
         }
L_100:
         qmin=qt[it-1][jt-1];
         }
      /* calculate the average tube wall temperature,thermal
      expansion,and clamped stress */
      if(jt==1||jt==NC)
         {
         et[it-1][jt-1]=0.;
         cs[it-1][jt-1]=0.;
         }
      else
         {
         tw[it-1][jt-1]=(tt[it-1][jt-1]/hai+ts[is-1][js-
      1]/hao)/(1./hai+1./hao);
         bdt=bt[ib-1]*(tw[it-1][jt-1]-tw0);
         et[it-1][jt-1]+=bdt*dlt*12.;
         cs[it-1][jt-1]=bdt*es[ib-1];
         }
      }
```

Appendix C: Monte Carlo Codes

In order to create a Monte Carlo model within Excel® two functions are necessary: one to return a normally distributed random number and a second to count the number of occurrences between a range of values. The following section of VBA® code provides these:

```
function random(mean As Double, std As Double) As Double
  Dim i As Integer, r As Double
  r = 0
  For i = 1 To 12
    r = r + Rnd()
  Next i
  random = mean + std * (r / 6 - 1)
End Function
Function CountBetween(X As Range, Xmin As Double, Xmax As Double) As Long
  Dim i As Long, n As Long
  n = X.Count
  CountBetween = 0
  For i = 1 To n
    If (X(i) >= Xmin) Then
      If (X(i) < Xmax) Then
        CountBetween = CountBetween + 1
      End If
    End If
  Next i
End Function
```

The following code implements the Monte Carlo heat exchanger model from Chapter 17 and easily handles one million cases.

```
#define _CRT_SECURE_NO_DEPRECATE
#include <stdio.h>
#include <stdlib.h>
#include <memory.h>
#include <math.h>

void*allocate(unsigned count,unsigned siz)
{
  void*ptr;
  if((ptr=calloc(count,siz))==NULL)
  {
    fprintf(stderr,"can't allocate memory\n");
    exit(1);
  }
  return(ptr);
}

int nint(double d)
{
```

```
    if(d>0.)
      return((int)(d+0.5));
    if(d<0.)
      return((int)(d-0.5));
    return(0);
    }

int urand()
    {
    int i,u;
    for(u=i=0;i<12;i++)
      u+=rand();
    return(u);
    }

double rnorm()
    {
    return(urand()/32767./6.-1.);
    }

double rdist(double a,double s)
    {
    return(a+6.*s*rnorm());
    }

double LMTD(double dT1,double dT2)
    {
    if(dT1<=0.||dT2<=0.)
      return(0.);
    if(fabs(dT1-dT2)<0.01)
      return(sqrt(dT1*dT2));
    return((dT1-dT2)/log(dT1/dT2));
    }

typedef struct{double avg,std;}RAN;

RAN mH ={4.00,0.40}; /* hot  side mass flow rate [kg/s]
    */
RAN CpH={3.50,0.05}; /* hot  side specific heat
    [kJ/kg/øC] */
RAN mC ={9.00,0.90}; /* cold side mass flow rate [kg/s]
    */
RAN CpC={2.50,0.04}; /* cold side specific heat
    [kJ/kg/øC] */
RAN THi={500.,0.67}; /* hot  side inlet temperature [øC]
    */
RAN THo={300.,0.67}; /* hot  side exit  temperature [øC]
    */
```

```
RAN TCi={250.,0.50};    /* cold side inlet temperature [øC]
*/
RAN TCo={375.,0.50};    /* cold side exit  temperature [øC]
*/
double Area=100.;       /* surface area [mý] */

#define value(x) rdist(x.avg,x.std)

typedef struct{double
   mh,cph,mc,cpc,thi,tho,tci,tco,dt,qh,qc,uh,uc;}RES;

RES heatx(int lst)
  {
  static RES res;
  res.thi=value(THi);
  res.tho=value(THo);
  res.tci=value(TCi);
  res.tco=value(TCo);
  res.mh =value(mH);
  res.mc =value(mC);
  res.cph=value(CpH);
  res.cpc=value(CpC);
  res.qh=res.mh*res.cph*(res.thi-res.tho);
  res.qc=res.mc*res.cpc*(res.tco-res.tci);
  res.dt=LMTD(res.thi-res.tco,res.tho-res.tci);
  res.uh=1000.*res.qh/Area/res.dt;
  res.uc=1000.*res.qc/Area/res.dt;
  if(lst)
    printf("%4.0lf %4.0lf %5.2lf %5.1lf
    %5.1lf\n",res.qh,res.qc,res.dt,res.uh,res.uc);
  return(res);
  }

typedef struct{int p[50];double a,s,x[51];}STA;

STA stats(double*X,int n,char*name)
  {
  int i,j,m;
  double A,Xm,S,Xx;
  static STA sta;
  memset(&sta,0,sizeof(sta));
  Xm=Xx=X[0];
  A=S=0.;
  for(i=0;i<n;i++)
    {
    if(X[i]<Xm)
      Xm=X[i];
    if(X[i]>Xx)
      Xx=X[i];
```

```
      A+=X[i];
      S+=X[i]*X[i];
      }
   A/=n;
   S=sqrt((S-n*A*A)/(n-1));
   printf("%s: min=%1G, avg=%1G, max=%1G,
      std=%1G\n",name,Xm,A,Xx,S);
   sta.a=A;
   sta.s=S;
   m=sizeof(sta.x)/sizeof(sta.x[0]);
   for(i=0;i<m;i++)
      sta.x[i]=Xm+i*(Xx-Xm)/(m-1);
   for(i=0;i<n;i++)
      {
      j=nint((X[i]-Xm)*((double)(m-1))/(Xx-Xm));
      if(j<0)
         j=0;
      else if(j>m-2)
         j=m-2;
      sta.p[j]++;
      }
   return(sta);
   }

int main(int argc,char**argv,char**envp)
   {
   char fname[]="monte.csv";
   int i,m,n=1000000;
   double*Qh,*Qc,*dT,*Uh,*Uc;
   FILE*fp;
   RES res;
   STA sQh,sQc,sdT,sUh,sUc;
   printf("Monte Carlo Heat Exchanger Model\n");
   printf("%i cases\n",n);
   Qh=allocate(n,sizeof(double));
   Qc=allocate(n,sizeof(double));
   dT=allocate(n,sizeof(double));
   Uh=allocate(n,sizeof(double));
   Uc=allocate(n,sizeof(double));
   for(i=0;i<n;i++)
      {
      res=heatx((i%100000)==0);
      Qh[i]=res.qh;
      Qc[i]=res.qc;
      dT[i]=res.dt;
      Uh[i]=res.uh;
      Uc[i]=res.uc;
      }
   printf("done\n");
```

```
    sQh=stats(Qh,n,"Qh");
    sQc=stats(Qc,n,"Qc");
    sdT=stats(dT,n,"dT");
    sUh=stats(Uh,n,"Uh");
    sUc=stats(Uc,n,"Uc");
    if((fp=fopen(fname,"wt"))==NULL)
      {
      fprintf(stderr,"can't create output file
    %s\n",fname);
      exit(1);
      }
    fprintf(fp,"Qh,pQh,Qc,pQc,dT,pdT,Uh,pUh,Uc,pUc\n");
    m=sizeof(sQh.x)/sizeof(sQh.x[0]);
    for(i=0;i<m-1;i++)
      {
      fprintf(fp,"%lG,%lG,"
    ,(sQh.x[i]+sQh.x[i+1])/2.,((double)sQh.p[i])/((double
    )n));
      fprintf(fp,"%lG,%lG,"
    ,(sQc.x[i]+sQc.x[i+1])/2.,((double)sQc.p[i])/((double
    )n));
      fprintf(fp,"%lG,%lG,"
    ,(sdT.x[i]+sdT.x[i+1])/2.,((double)sdT.p[i])/((double
    )n));
      fprintf(fp,"%lG,%lG,"
    ,(sUh.x[i]+sUh.x[i+1])/2.,((double)sUh.p[i])/((double
    )n));

      fprintf(fp,"%lG,%lG\n",(sUc.x[i]+sUc.x[i+1])/2.,((dou
    ble)sUc.p[i])/((double)n));
      }
    fclose(fp);
    printf("see %s for results\n",fname);
    return(0);
    }
```

The output looks like this:

```
Monte Carlo Heat Exchanger Model
1000000 cases
3164  2701  81.09  390.1  333.1
2970  2953  82.39  360.5  358.4
2630  2680  81.65  322.1  328.3
3106  2940  81.99  378.9  358.6
2559  3091  81.63  313.5  378.7
3128  2953  81.56  383.5  362.1
3271  3303  82.19  398.0  401.9
2357  2451  81.42  289.5  301.0
2646  2611  82.02  322.6  318.4
2916  3186  81.99  355.7  388.6
done
```

```
Qh: min=1546.8,   avg=2800.4,   max=4099.52,  std=283.679
Qc: min=1576.23,  avg=2812.87,  max=4008.61,  std=285.542
dT: min=78.5398,  avg=81.8496,  max=84.9284,  std=0.660767
Uh: min=191.825,  avg=342.165,  max=496.741,  std=34.8006
Uc: min=191.586,  avg=343.683,  max=494.223,  std=34.9743
see monte.csv for results
```

The graphics are in the second section of Chapter 17.

also by D. James Benton

3D Articulation: Using OpenGL, ISBN-9798596362480, Amazon, 2021 (book 3 in the 3D series).
3D Models in Motion Using OpenGL, ISBN-9798652987701, Amazon, 2020 (book 2 in the 3D series.
3D Rendering in Windows: How to display three-dimensional objects in Windows with and without OpenGL, ISBN-9781520339610, Amazon, 2016 (book 1 in the 3D series).
A Synergy of Short Stories: The whole may be greater than the sum of the parts, ISBN-9781520340319, Amazon, 2016.
Azeotropes: Behavior and Application, ISBN-9798609748997, Amazon, 2020.
bat-Elohim: Book 3 in the Little Star Trilogy, ISBN-9781686148682, Amazon, 2019.
Boilers: Performance and Testing, ISBN: 9798789062517, Amazon 2021.
Combined 3D Rendering Series: 3D Rendering in Windows®, 3D Models in Motion, and 3D Articulation, ISBN-9798484417032, Amazon, 2021.
Complex Variables: Practical Applications, ISBN-9781794250437, Amazon, 2019.
Compression & Encryption: Algorithms & Software, ISBN-9781081008826, Amazon, 2019.
Computational Fluid Dynamics: an Overview of Methods, ISBN-9781672393775, Amazon, 2019.
Computer Simulation of Power Systems: Programming Strategies and Practical Examples, ISBN-9781696218184, Amazon, 2019.
Contaminant Transport: A Numerical Approach, ISBN-9798461733216, Amazon, 2021.
CPUnleashed! Tapping Processor Speed, ISBN-9798421420361, Amazon, 2022.
Curve-Fitting: The Science and Art of Approximation, ISBN-9781520339542, Amazon, 2016.
Death by Tie: It was the best of ties. It was the worst of ties. It's what got him killed., ISBN-9798398745931, Amazon, 2023.
Differential Equations: Numerical Methods for Solving, ISBN-9781983004162, Amazon, 2018.
Equations of State: A Graphical Comparison, ISBN-9798843139520, Amazon, 2022.
Evaporative Cooling: The Science of Beating the Heat, ISBN-9781520913346, Amazon, 2017.
Forecasting: Extrapolation and Projection, ISBN-9798394019494, Amazon 2023.
Heat Engines: Thermodynamics, Cycles, & Performance Curves, ISBN-9798486886836, Amazon, 2021.
Heat Recovery Steam Generators: Thermal Design and Testing, ISBN-9781691029365, Amazon, 2019.

Heat Transfer: Heat Exchangers, Heat Recovery Steam Generators, & Cooling Towers, ISBN-9798487417831, Amazon, 2021.
Heat Transfer Examples: Practical Problems Solved, ISBN-9798390610763, Amazon, 2023.
The Kick-Start Murders: Visualize revenge, ISBN-9798759083375, Amazon, 2021.
Jamie2: Innocence is easily lost and cannot be restored, ISBN-9781520339375, Amazon, 2016-18.
Kyle Cooper Mysteries: Kick Start, Monte Carlo, and Waterfront Murders, ISBN-9798829365943, Amazon, 2022.
The Last Seraph: Sequel to Little Star, ISBN-9781726802253, Amazon, 2018.
Little Star: God doesn't do things the way we expect Him to. He's better than that! ISBN-9781520338903, Amazon, 2015-17.
Living Math: Seeing mathematics in every day life (and appreciating it more too), ISBN-9781520336992, Amazon, 2016.
Lost Cause: If only history could be changed..., ISBN-9781521173770, Amazon, 2017.
Mass Transfer: Diffusion & Convection, ISBN-9798702403106, Amazon, 2021.
Mill Town Destiny: The Hand of Providence brought them together to rescue the mill, the town, and each other, ISBN-9781520864679, Amazon, 2017.
Monte Carlo Murders: Who Killed Who and Why, ISBN-9798829341848, Amazon, 2022.
Monte Carlo Simulation: The Art of Random Process Characterization, ISBN-9781980577874, Amazon, 2018.
Nonlinear Equations: Numerical Methods for Solving, ISBN-9781717767318, Amazon, 2018.
Numerical Calculus: Differentiation and Integration, ISBN-9781980680901, Amazon, 2018.
Numerical Methods: Nonlinear Equations, Numerical Calculus, & Differential Equations, ISBN-9798486246845, Amazon, 2021.
Orthogonal Functions: The Many Uses of, ISBN-9781719876162, Amazon, 2018.
Overwhelming Evidence: A Pilgrimage, ISBN-9798515642211, Amazon, 2021.
Particle Tracking: Computational Strategies and Diverse Examples, ISBN-9781692512651, Amazon, 2019.
Plumes: Delineation & Transport, ISBN-9781702292771, Amazon, 2019.
Power Plant Performance Curves: for Testing and Dispatch, ISBN-9798640192698, Amazon, 2020.
Practical Linear Algebra: Principles & Software, ISBN-9798860910584, Amazon, 2023.
Props, Fans, & Pumps: Design & Performance, ISBN-9798645391195, Amazon, 2020.
Remediation: Contaminant Transport, Particle Tracking, & Plumes, ISBN-9798485651190, Amazon, 2021.

ROFL: Rolling on the Floor Laughing, ISBN-9781973300007, Amazon, 2017.
Seminole Rain: You don't choose destiny. It chooses you, ISBN-9798668502196, Amazon, 2020.
Septillionth: 1 in 10^{24}, ISBN-9798410762472, Amazon, 2022.
Software Development: Targeted Applications, ISBN-9798850653989, Amazon, 2023.
Software Recipes: Proven Tools, ISBN-9798815229556, Amazon, 2022.
Steam 2020: to 150 GPa and 6000 K, ISBN-9798634643830, Amazon, 2020.
Thermochemical Reactions: Numerical Solutions, ISBN-9781073417872, Amazon, 2019.
Thermodynamic and Transport Properties of Fluids, ISBN-9781092120845, Amazon, 2019.
Thermodynamic Cycles: Effective Modeling Strategies for Software Development, ISBN-9781070934372, Amazon, 2019.
Thermodynamics - Theory & Practice: The science of energy and power, ISBN-9781520339795, Amazon, 2016.
Version-Independent Programming: Code Development Guidelines for the Windows® Operating System, ISBN-9781520339146, Amazon, 2016.
The Waterfront Murders: As you sow, so shall you reap, ISBN-9798611314500, Amazon, 2020.
Weather Data: Where To Get It and How To Process It, ISBN-9798868037894, Amazon, 2023.

www.ingramcontent.com/pod-product-compliance
Lightning Source LLC
Chambersburg PA
CBHW031430210526
45464CB00005B/2139